HSU PRESS

ideaFest Journal

Volume 4 | 2020

The Interdisciplinary Journal of
Creative Works and Research
from Humboldt State University

Humboldt State University Press

Humboldt State University Library
1 Harpst St.
Arcata, CA 95524-8299

digitalcommons.humboldt.edu

ISBN: 978-1-947112-48-3

Managing Editor: Aaron Laughlin
Design by Aaron Laughlin

Humboldt State University Press publishes high-quality scholarly, intellectual, and creative works by or in support of our campus community. The press supports the HSU mission to improve the human condition and our environment by promoting understanding of social, economic, and environmental issues.

ideaFest Contents Journal

Anthropology

The Biocultural Trauma Feedback Loop — 4
Michelle Irvine

Biological Sciences

17 — Influence of Pinus contorta and Arctostaphylos uva-ursi on the presence of Cladonia portentosa subsp. pacifica within protected dune ecosystems in Arcata, California
Courtney Michelle Harris, Kacia Layman, Jamie Wautlet, Grace Rhoades, Chris Belko

23 — Analysis of the Variables Affecting Plant Species Richness in Deserts
Eli Kallison, Ellen Thompson, Maddison Keen, Rusty Newman

32 — Exploration of Antiviral Effects of RNA-dependent RNA Polymerases 3, 4 and 5 in Arabidopsis
Eli Kallison, Aleksandra Berici, Blake C. Meyers

38 — Differences in algal, gastropod, and arthropod coverage surrounding Pisaster ochraceus and other asteroids in Humboldt County, California
Desa Smith, Jana Lott, Savannah Williams, Shayan Farzadpour

Chemistry

Preliminary Quantitative and Qualitative Investigations on the Heterogeneous, Irreversible Uptake of Acetone by Titanium(IV) Oxide — 44
John S. André, Natalie Eskandary, Terry Franklin, Emilia J. McCann, Oscar Pascual-Diaz, Gregory M. Pitch, Jesus D. Rincon, Megan N. Wright, Christopher W. Harmon

English Studies

59 — Levine's Marxist Toys
Dean Engle

Native American Studies

Reclaiming Indigenous Women's Roles in the 21st Century — 65
Crystal Miller

Physics

72 — Scaling Resistance with Channel Length for Carbon Nanotube Networks
Cade Freels, Carla Patricia Quintero, Samantha A. Anger, Will Gannett, Ruth Saunders

78 — Percolation in Carbon Nanotube Networks
William P. Grieder, Will Gannett, Ruth Saunders

The Biocultural Trauma Feedback Loop

Michelle Irvine

Abstract

It is widely known that trauma is repeated throughout a victim's life, but the biological mechanisms of its recurrence (revictimization), even though understood biologically, are not accepted or discussed in all disciplines. A combination of socio-cultural and biological perspectives is needed to understand this cycle of revictimization and to offer help for sufferers and public health agencies. In order to better understand these issues, I conducted a synthesis of existing scientific research regarding the discrepancies between biological and sociological studies on revictimization. Within my review of sociological research it was revealed that initial trauma and revictimization are clearly understood as a positive feedback loop, with one increasing the other over a victim's life. In biology, however, this loop has been acknowledged but the study of the recurrence of trauma has not been integrated into these disciplines. In humans, biology and sociology are inseparable; a recognition in these disciplines that a positive feedback loop exists regarding revictimization is key. Recognizing the existence of this biological feedback loop has the potential to mitigate the damage of past, present, and future trauma. With a better understanding of the biological aspects of their recurring trauma, devastated sufferers can be empowered against damaging ideologies, such as biological determinism and victim blaming.

Keywords: revictimization, trauma, positive feedback loop

Introduction

The experience of a traumatic event is all encompassing. It forever alters an individual's genes, how those genes are expressed, and the synthesis of proteins and other molecules. Those alterations affect how organ systems, such as the nervous and cardiovascular, operate. The function of organ systems affects the body as a whole. The body affects the mental and emotional states of an individual. Mental and emotional functioning changes how societies and cultures perceive said individuals. Trauma affects everything that a human being is and everyone surrounding them. No person is untouched by trauma.

It is widely known that trauma is repeated throughout a victim's life, but the biological mechanisms of its recurrence, even though understood biologically, are not accepted or discussed in all disciplines. Within my review of existing biological research on trauma, there was severe compartmentalization, separating the work of different disciplines from the biological sciences, in addition to the removal of lived human experiences. The majority of the studies I reviewed couched their language toward trauma in blanket terms such as "stress". This term mutes survivors of trauma and equates their "stressful" experiences with the same "stress" that a student might endure during finals week. The depths of these studies are invaluable to trauma research as a whole; however, the language used to describe trauma (or stress) is at best problematic and at worst an open door to biological determinism, scientific victim blaming, and an overall reductionist view of the lived human experience. Sociological studies of trauma do not fare much better on their own, either.

Within my review of sociological research it was revealed that initial trauma and revictimization are clearly understood as a positive feedback loop, with one increasing the other over a victim's life. Unfortunately, this was where most articles ended their studies of trauma and began to decry the biological sciences for not integrating the "material body" with the body embedded in time, space, and experience, as described by Margaret Lock in her synthesis of recent epigenetic research (Lock, 2015). There was also the occurrence of sociological studies that reference biological sciences (specifically genetics and epigenetics) without a nuts-and-bolts understanding of the topic. This does not aid the outcry for future cooperative, interdisciplinary research between the sociological and biological sciences. It has become apparent that more must be done to bridge the divide.

A combination of socio-cultural and biological perspectives is needed to understand this cycle of revictimization and to offer help for sufferers and public health agencies. In order to accomplish this goal, sociological sciences must become more genetically and epigenetically literate. Future research into trauma and revictimization should include fully integrated, holistic perspectives so that the experiences of survivors are recognized by all disciplines.

In humans, biology and sociology are inseparable, a foundational tenant of anthropology that is not acknowledged by various biological fields of study. A recognition in the biological sciences that a positive feedback loop exists regarding revictimization, is key towards addressing the complexities of the issue of trauma. Conversely, recognizing and integrating the existence of the biological facet of the feedback loop has the potential to mitigate the damage of past, present, and future trauma. With a better understanding of the multifaceted aspects of their recurring trauma, devastated sufferers can be empowered against damaging ideologies, such as biological determinism and victim blaming.

Methodology

For this project, I conducted a review of existing research, focused on salient findings between the years 2000-2019 regarding the discrepancies between biological and sociological studies on revictimization. In my analysis of this research, I attempted to integrate findings from studies within these disciplines. The goal was to unify some of the research of these fields into a holistic, biocultural trauma feedback loop that stretches from the genetic to the sociological causes of revictimization (the reoccurance of trauma). In addition to outlining an integrated feedback loop, I created an extremely tentative example of the feedback loop in action. One gene was isolated and followed from its epigenetic modifications, molecular alterations, organ system changes, psychosocial and emotional impacts, all the way to sociocultural responses.

Sociological Feedback Loop

There are a multitude of avenues in which trauma can be studied. This section will focus primarily on childhood trauma and mental health issues as the path toward revictimization, keeping intersectionality, resilience, and the lived human experience in mind.

It is no secret that the experience of trauma causes damage to the survivors. When those survivors are children, with yet fully developed brains, the damage can be permanent. In a review of neurobiology research on childhood trauma utilizing the Adverse Childhood Experience Study as a case example, Anda et. al found that: "...early life stress such as abuse and related adverse experiences cause enduring brain dysfunction that, in turn, affects health and quality of life throughout the lifespan" (Anda et. al, pg. 175). The study further explains that their findings support "the hypothesis of dysfunction", which states that early childhood trauma alters the functioning of particular regions of the brain that are currently believed to be involved in anxiety and mood regulation. In another study, which conducted a statistical and systematic analysis of childhood trauma (or maltreatment) types and negative mental health outcomes, it was found that all the types of maltreatment under question had a statistically significant relationship with poor mental health outcomes (Cecil et. al, pg. 114). Childhood trauma, maltreatment, or adverse events all appear to have similar effects on the developing brain: dysfunction occurs, mental health issues ensue, and the quality of a child's life decreases. To further exacerbate the situation, children develop negative coping skills in response to these experiences.

In the large-scale, nationally-representative, epidemiological analysis of the United States conducted by Vaughn-Coaxum et. al, it was found that: "... exposure to nearly all forms of trauma was... strongly associated with increased negative emotion-focused coping" (Vaughn-Coaxum et. al, pg. 842). The use of the coping styles that were under observation, only increased as an individual experienced a greater number of traumatic events. Furthermore, their results suggested that forms of coping linked to poor mental health were related to the amount of traumatic events an individual had prior to participation in the study (Vaughn-Coaxum et. al, pg. 850). The World Health Organization, in a study on violence and mental health that utilized more than 160 experts from over 70 countries, also found that traumatic experiences with violence have a cumulative effect on the mental health of victims. The mental health of victims is greatly impacted by the severity, type, and number of experiences with violence that occur throughout their lifetimes. The greater the severity, type, and number of occurrences, the more negative mental health outcomes can ensue (Mercy et. al, pg. 22). The World Health Organization, noticed that poor coping behaviors, such as risk taking, were likely caused by childhood exposure to trauma or maltreatment (Mercy et. al, pg. 23). There is no singular cause that makes individuals more likely to become a victim of violence; however, there is a relationship between childhood mental health disorders, trauma, and negative life trajectories.

In a fifteen-year-long study of 3,804 schoolchildren in Hong Kong, conducted by Raine et. al, that was concerned with peer victimization and children with schizophrenia or schizotypal symptoms, it was noted that adult individuals with the disorder are: "...14 times more likely to be the victims of a violent crime" (Raine et. al, pg. 938). This is in addition to their discovery that the symptomatology of schizophrenia in school-aged youth causes other children to treat them poorly (Raine et. al, pg. 937). Schizophrenia and its spectrum are all found within a larger grouping of mental health disorders, specifically the "trauma spectrum". This spectrum includes: anxiety and panic disorders, depression, schizophrenia and other dissociative disorders. As previously discussed, the Anda et. al study goes into detail describing the comorbidity of these disorders. Hence the term "trauma spectrum disorders", which describes how they all occur together in different combinations and levels of severity (Anda et. al, pg.182). Mental health disorders, according to the World Health Organization study, are "both causes and consequences of interpersonal, collective, and self-directed violence" (Mercy et. al, pg. 20).

Copeland et. al found in their review of the Great Smoky Mountain Study, in which 1,420 children were interviewed through adolescence and young adulthood about their experiences with bullying, that this childhood event is much more serious than previously believed. To experience bullying in childhood is known to have negative impacts on development; however, recent research suggests that bullying can be as severe as maltreatment from family members (Copeland et. al, pg. 7573). A potential basis for bullying and other hierarchical behavior among children was also discovered through this study. Copeland et. al noticed that children who were purely bullies had significantly less low grade inflammation throughout their lives, while pure victims and bully-victims had the most low grade inflammation (compared to children who were not involved in bullying at all). Long term low grade inflammation is associated with high levels of depressive disorders and other health risks (Copeland et. al, pg. 7570). Victimizing others gives an individual biological advantages (Copeland et. al, pg. 7573). Populations that are already vulnerable are the most at risk, such as: minorities, refugees, those with disabilities, the elderly, women, the LGBTQA+ community, and many others (Mercy et. al, pg. 28).

The World Health Organization study, conducted by Mercy et. al, outlined a multiplicity of risk factors for experiencing violence on the individual level. "Individual-level risks identified include demographic factors such as age, sex, and race/ethnicity; psychological and personanlity disorders, alcohol and substance abuse, and a history of engaging in violent behavior or experiencing abuse" (Mercy et. al, pg. 28). Taken together, it becomes apparent that disenfranchised communities are hit the hardest and held the longest by the feedback loop of trauma. Ethnicity and socioeconomic status, in this particular example and social climate, cause a cascade of psychosocial and mental-emotional issues that are compounded by victimization and structural violence. This experience has been named "racial battle fatigue". In the chapter titled, "The Epigenetics of Being Black and Feeling Blue", in which a synthesis of preexsisting research is utilized, racial battle fatigue is described as: "...a state of keeping the bodies of stigmatized minorities hypervigilant in anticipation of the next white racial insult. Racial battle fatigue affects human biology and physiology at the cellular level, leaving the bodies of the poor, the impoverished, and the targeted more vulnerable to mental and physical health decline" (Smith, pg. 261). Smith goes on to explain how disenfranchised communities also struggle with lifestyle related disease. In order to cope with hypervigilance and the sensitivity that follows, risky lifestyles are often adopted by disenfranchised populations to relieve some of their burdens (Smith, pg. 262). Keeping in mind the Copeland et. al study on childhood bullying, there is a biological boost for those who wish to keep bullying minorities and other disenfranchised communities. There is, however, one defense that these communities have: resilience.

The chapter titled "Impact of Trauma: Vulnerability and Resilience", which relied on professional experience and a review of scientific literature, examined the recovery of victims of genocidal trauma using a perspective that recognizes the coexistence of strength and vulnerability inside each survivor (Giberovitch and Barry, pg. 68). Recovering from trauma is a difficult and ill-defined task. It never looks quite the same for any individual. The chapter outlined a few of the tasks that are necessary for recovery to begin. "Healing can be difficult because it requires accomplishing psychic tasks such as creating a sense of rootedness, belonging, and continuity with the past; working through losses, guilt, rage, and shame; and meaningfully integrating the experience into the totality of life" (Giberovitch and Barry, pg. 69). They go on to explain that the magnitude of the Holocaust may not be able to be fully integrated into an individual's life, that it will take more than a single generation or lifetime. The same can most certainly be said of the historic trauma of slavery. The key to assisting survivors toward whatever kind of recovery they need is empathy. One author observed that the types of therapy they utilized often mattered less than listening empathetically to survivors and creating honest, supportive relationships with them (Giberovitch and Barry, pg. 83). The chapter goes on to explain that survivors also fare better when they see actual biological proof that validates their experiences. When discussing experiences, such as intrusive memories, with their clients, one author noted the empowerment that survivors feel when their personal experiences are validated by research, as well as the comfort of knowing that other survivors are having similar experiences (Giberovitch and Barry, pg. 83). This validation can also help the families of survivors to be more sensitive and empathetic towards the symptomatology of trauma. Strong, caring, and empathetic relationships are the keys to recovery and are the foundations on which resilience, even in early childhood, is developed. In his article on early childhood development, Jack Shonkoff, from the Center on the Developing Child at Harvard University, discussed the source from which children develop their ability to cope with hardships through the synthesizing of preexisting research. It is through supportive relationships and chances to develop positive coping mechanisms, that children acquire the ability to face adversity (Shonkoff, pg. 12). From childhood traumas to historical trauma, survivors need to be treated with empathy, on an individual basis, be involved in close, caring relationships, and have their experiences validated by science. This integrative view of survivor resilience requires a biocultural approach; however, trauma and revictimization are not solely the ailments of individuals, but of whole societies. To rephrase a quote from James Baldwin: Trauma is stuck in us, and we are stuck in trauma.

Epigenetics: Between Biology and Sociology

In this section, the primary focus will be on the epigenetic linkage of trauma and revictimization, including their sociological and biological aspects. This will be in addition to outlining how epigenetic modifications from trauma generally occur. Due to the nature and current capability of epigenetic research, a portion of cited research in this section can only theorize how their non-human based findings can potentially be applied to future human studies.

Genes are not static and unchanging. They wait for signals from the external world, such as the environment, food resources, and the social and physical location of our homes. Genes also receive signals from the sociological aspects of our lives, like socio-economic status and disenfranchisement (Smith, pg. 261). The experience of trauma leaves markers on the genes of victims, modifying how DNA is transcribed and replicated. This occurs through various processes such as histone modification, DNA methylation, mitochondrial modification, and many other avenues. A study in which mitochondrial functioning was selectively dysregulated in mice and then systematically characterized, it was found that mitochondrial defects accounted for 81 percent of the observed, hippocampal gene expression changes that were associated with alterations to the brain's stress response system (Picard et. al, pg. E6619). Their study sought to uncover if there was a relationship between defects in the stress response systems of the body and dysregulation of mitochondrial gene expression. Therefore, their research operated under the notion that: "...by supplying the majority of cellular energy, mitochondria contribute to the organism's overall adaptive capacity" (Picard et. al, pg. E6621). In another article that examined genome wide association studies in order to elucidate new genes involved in the development of PTSD, the role of mitochondria was also explored. Recent research, focused on the mitochondria, found that shifts in the membrane potential of the mitochondrian may be a potential cause of the reduction in hippocampal volume that is characteristic of PTSD sufferers (Feodorova and Sarafian, pg. 9). In other words, the mitochondria affected genetic and non-genetic signalling to glucocorticoid receptors (GR) within the hippocampus, which in turn affect various transcription factors that control the expression of thousands of genes (Feodorova and Sarafian, pg. 6). Taken together, the sociological causes of trauma cascade into the epigenetic precursors of revictimization. To again cite the World Health Organization study, the experience of violence, and its effects on one's mental health, are cumulative, with the incidence of one increasing the incidence of the other (Mercy et. al, pg. 22). It can then be suggested that, as an individual experiences more trauma, more epigenetic markers can be left on their genome, which causes more changes in gene expression to accumulate.

One mechanism in which epigenetic modifications are accumulated is through histone modification. Histone modifications are much more plastic and less stable than DNA methylation. They can be placed on a particular region of the genome, taken off, and placed back (Carey, pg. 72), as discussed by Nessa Cary in her book *The Epigenetics Revolution*. She is a professor at Imperial College in London and has specialized in epigenetics for eighteen years. This plasticity allows cells to experiment with various gene expression patterns. "If there is an advantage to the cell in those genes being switched off, the histone modifications may last long enough to lead to DNA methylation" (Carey, pg. 72). Once DNA becomes methylated, it tends to stay that way, making the modified gene expression heritable (Carey, pg. 72). There is, however, variation in how this expression plays out. This is due in part to the proteins that "read" histone modifications. They bring out other proteins, which combine with the "readers" into complex structures that can turn certain gene expressions on and off (Carey, pg.73). Depending on individual protein availability, some of the epigenetic markers of trauma can be either turned on or off. As Catherine Malabou discussed in her synthesis of existing research, the experience of trauma is different for everyone; some individuals feel their trauma more acutely than others, some trauma is inherited rather than experienced first hand, while others appear unaltered by it or do not remember (Malabou, pg. 187). Dr. Howe, a professor at Humboldt State University and the chair of the Psychology department, discussed the multifaceted nature of where these variations, in response to trauma, arise. The HPA axis, which regulates stress response, functions differently between individuals. This results in various reactions to trauma based on individual experiences, brain chemistry, genotypes, coping skills, all within the specific zeitgeist in which we live (Howe, pg. 472).

All of these factors are in addition to systems of power, privilege, and oppression, in which disenfranchised individuals and communities are left more vulnerable to negative environmental and genetic outcomes. Joseph Graves stated in his synthesis of preexisting research that: "Epigenetic programming from past experience (or inheritance) can enhance vulnerability to stress and place the brain at higher risk of neurodegenerative events. Stress, defined as a perturbation of homeostasis, can cause the onset of complex neural and endocrine responses characterized by activation of the HPA axis" (Graves, pg. 39). Activation of the HPA axis is directly associated with stress; dysregulation of the HPA axis is associated with trauma and trauma is a neurodegenerative event. The HPA axis influences the hippocampus and the amygdala, which are affected by trauma due to their regulation of memory and fear (Giberovitch and Barry, pg. 80). One particular gene, the ADRA2b gene, is associated with the dysregulation of brain regions associated with the HPA axis. When an individual carries a particular deletion mutation of the ADRA2b gene, said individual is significantly more reactive to emotional stimuli (Todd et. al, pg. 6514). The occurrence of trauma causes the hippocampus to shrink, due to hyperactivity in the amygdala. In a study that utilized magnetic resonance imaging on neurosurgical patients with ventromedial prefrontal cortex damage, it was found that hyperactivity in the amygdala is characterized by unregulated reactivity to negative, emotional stimuli (Motzkin et. al, pg. 276). A traumatic experience, therefore, is a prerequisite for future victimization and/or PTSD; however, not all individuals who experience trauma are genetically predisposed to these outcomes. Research on twins suggests that approximately 30 percent of the variation in PTSD symptomology is due to genetic variation (Feodorova and Sarafian, pg. 9). All of the gene expression changes from trauma are heritable; they are passed on through the generations regardless of whether or not they become activated (or deactivated) during any particular individual's lifetime (Carey, pg.73). This is emblematic of historical trauma like that of Holocaust survivors and African American communities, as well as intergenerational traumas of violence and incest. The legacy of trauma is remembered by the genes and the body, regardless of the passage of time. This causes the bodies of the disenfranchised to remain hypervigilant. It is the larger society, and it's views on trauma and victims, that interacts with susceptible genomes. As Darron Smith explained: "This dynamic interaction between our genes and environment should serve as a reminder that human beings thrive in the absence of war, famine, and other manufactured discord" (Smith pg. 262).

Biological Research

This section will discuss the biological mechanisms of trauma and revictimization. A multiplicity of research from various disciplines within the biological sciences will be utilized.

A promising new topic of inquiry, into the stress response systems of the body, is the role of mitochondria and its DNA. The conclusions of the Picard et. al study coincided with these recent revelations. "This result is consistent with recent findings that mitochondrial dysfunction can exert robust and bidirectional regulation across the majority (~70%) of genes within the human genome" (Picard et. al, pg. E6622). In another study conducted by Picard et. al, mitochondrial functions were selectively manipulated in order to observe potential changes to stress response systems. Genetic and epigenetic modifications are all induced through the activation of metabolic pathways that are localized in the mitochondria. "Thus, both the addition and removal of epigenetic marks are metabolically—or mitochondrially—regulated" (Picard et. al, 73). The mutations and defects of the mitochondria and its DNA, due to its energetic and regulatory functions, can affect the transcription and replication of nuclear DNA. These defects are much more common in mitochondrial DNA than in nuclear DNA. There are approximately two to fifteen copies of mitochondrial DNA within a single mitochondrian that replicate once each month. This creates a higher mutation rate in mtDNA than in nuclear DNA, in addition to mitochondrial cell division which occurs independently of mitosis, as discussed in the Sas et. al review of existing scientific literature (Sas et. al, pg. 3). Thus, the vulnerability of the mitochondrial DNA (mtDNA) becomes the vulnerability of the nuclear DNA (nDNA). It has been theorized that oxidative stress is the culprit. This is due to the fact that mitochondria create the majority of reactive oxygen species. Due to the derived nature of mitochondria, each one still contains circular DNA, a hallmark of their bacterial origins. This is their individual, mitochondrial DNA in which key components of the oxidative phosphorylation process are contained, which Pinti et. al explained in their review of current cellular biology research (Pinti et. al, pg. 1). These studies point to the evolutionary history of the stress response system within multicellular life and the beneficial relationship between eukaryotes and mitochondrial ancestors.

Epigenetic modifications to gene expression, due to stress or trauma, occur in a multiplicity of regions within the body. Specifically, direct epigenetic changes occur in the expression of genes associated with the HPA axis, a particular region of the brain that regulates stress, fear, and emotion; it is the hub of the brain's stress response system. Some of the defects in the brain's stress response systems that are specifically associated with trauma are: a shrinkage in the hippocampus, an overactive amygdala, and cortisol levels becoming dysregulated. These factors were noted in the Feodorova and Sarafian study. "Decreased activity and neuronal atrophy in the hippocampus and in the PFC, as well as increased activity and neuronal growth in the amygdala have been confirmed" (Feodorova and Sarafian, pg. 6). They also noted the relationship between changes to the hippocampus and cortisol. The vulnerability of the hippocampus to stress is due to its high number of corticosteriod receptors (Feodorova and Sarafian, pg. 6). These receptors are responsible for the secretion and the reuptake of these cortisol-based hormones. Too much secretion of cortisol can lead to immunosuppression, while too little reuptake can lead to autoimmune disorders; both outcomes lead to poor mental health. In a study examining the outcome of forced swim (FS) in mice, which later extrapolates to future human research, it was found and hypothesized that: dysregulation of glucocorticoid secretion, due to trauma and/or chronic stress, increases an individual's vulnerability to mental health disorders (Mifsud and Reul, pg. 11336). The study goes on to explain that: "Glucocorticoids act via mineralocorticoid receptors (MRs) and glucocorticoid receptors (GRs) in the hippocampus, resulting in altered transcription of target genes" (Mifsud and Reul, pg. 11336). With defects in the transcription of its genes, the hippocampus cannot go through neurogenesis properly, hence the neuronal atrophy. This has the potential to explain the difficulties with memory that are associated with traumatic events, since the hippocampus plays a vital role in the formation of long term memories.

The amygdala, on the other hand, is associated with the processing of memories, as well as emotional responses to stressful stimuli. When working properly, the amygdala assists in limiting responses to stress. After trauma, the amygdala becomes dysregulated and intense, emotional reactions to stimuli ensue. This is due to the previously discussed gene, ADRA2b. A study, in which fMRIs were taken in order to conduct neural pathway analyses, found that "...emotionally salient stimuli are subjectively experienced with greater perceptual vividness, a phenomenon we call emotionally enhanced vividness (EEV). EEV has been linked to greater activation of object-sensitive regions of the visual cortex, an effect mediated by amygdala activity" (Todd et. al, pg. 6506). The effects of a smaller hippocampus and an overactive amygdala can also cause an increase in the likelihood of developing poor coping behaviors, due to the shrinkage of the hippocampus and the anterior cingulate cortex (ACC), which regulates rational decision making. The regulation of rational decision making by the ACC and hippocampus, is itself regulated by hormone secretion, namely cortisol. When cortisol levels are dysregulated, inflammation occurs due to cortisol's (usually) anti-inflammatory properties. Increased and prolonged inflammation is associated with the development of mental health disorders (Copeland et. al, pg. 7570). HPA axis functioning and cortisol regulation reported in victims of bullying, were consistent with the researchers' data illustrating increases in CRP (C-reactive protein, which is a marker of low-grade systemic inflammation) levels of said victims (Copeland et. al, pg. 7572). The hormonal, neurological, and inflammatory responses to stress exist in a positive feedback loop for those susceptible. Susceptibility and resilience are two sides of the same proverbial coin.

In one study, titled "Blood-Brain Biomarkers for Stress Susceptibility", Chattarji and Rao reviewed human and non-human neurobiology research and discovered several genes that may play into individual variation of stress susceptibility. "Among these nine factors, the glucocorticoid receptor (NR3C1) is of particular interest as the development of PTSD-like symptoms can be blocked by the administration of glucocorticoids shortly before or after stress" (Chattarji and Rao, pg. 13254). Their study also found that the particular genes under question were associated with both the amygdala and the hippocampus. These regions of the brain, and their dysregulation, are clearly understood to be involved in the development of PTSD (Chattarji and Rao, pg. 13254). Susceptibility to the negative effects of trauma stem from genetic variations downstream and the dysregulation of specific brain regions upstream.

This susceptibility is due to a bias against long-term neurological needs, in favor of short term protection. Recent research has shown that an evolutionarily salient mechanism favors habitual, stress reduction behaviors over executive functioning during heightened stress in order to promote short term survival (Sinha et. al, pg. 8837). This may explain why the hippocampus and the amygdala become dysregulated, due to their association with the prefrontal cortex, which controls executive functioning.

In their study on resilient coping, utilizing an exposure paradigm and fMRI, Sinha et. al discovered that "...greater neural flexibility signals in the ventromedial prefrontal cortex during stress correlated with active coping ratings whereas lower dynamic activity in the VmPFC also predicted a higher level of maladaptive coping behaviors in real life" (Sinha et. al, pg. 8837). Due to trauma's effect on the prefrontal cortex, decision making and stress-response regulation are much more difficult for those affected by a multitude of traumatic events over their lifetimes, hence the maladaptive coping behaviors. This is in addition to the increase in vulnerability to mental health disorders (Vaughn-Coaxum et. al, pg. 843). Long term, low-grade inflammation, a marker of chronic stress, is another factor that increases one's vulnerability to mental health disorders, such as PTSD (Copeland et. al, pg. 7570).

In another study on resilience, in which mice were repeatedly inoculated with M. vaccae against stress related inflammation, a bacterial connection to the stress response system was uncovered. A reduction in T cells, increased IBD, elevated proinflamatory responses, and autoimmunity issues are all positively correlated to PTSD (Reber et. al, pg. E3130). The close relationship between the brain, stress, and the bowels of humans is well understood, as discussed by Moloney et. al in their review of current research on the microbiota-gut-brain axis (Moloney et. al, pg. 104); however, this particular study focused on animal models and bacterial treatments for gastrointestinal sources of stress.

Reber et. al, found that "...immunization with M. vaccae induced a long lasting shift toward a more proactive coping response, characterized by decreased submissive, flight, and avoiding behaviors, during chronic psychosocial stress that, based on previous studies in rodents and humans, may decrease vulnerability to the development of more persistent anxiety- and depressive-like symptoms" (Reber et. al, pg. E3131). There is undoubtedly a connection between the biome in the gut and the functioning of the mind. When synthesized, these studies suggest that genetic differences in the neural structuring of particular regions of the brain, act together with the microbiota of the gastrointestinal tract to cause susceptibility (or resilience) to stress.

Tentative Findings of a Biocultural Feedback Loop

In this section, all the previously discussed studies will be synthesized into a tentative biocultural trauma feedback loop. In particular, the relationships between trauma and revictimization will be elucidated through: childhood trauma, interpersonal violence, and mental health disorders.

At the genetic level, a deletion mutation occurs on the ADRA2b gene. This mutation results in carriers experiencing emotional stimuli in a more intense manner than non-carriers. Carriers of this deletion have more vivid and emotional perceptions of the world due to increased activity in regions of the brain associated with evaluation. Differences in serotonin and norepnephrine availability, as well as personal experiences, dictate what our brains view as emotionally valuable input (Todd et. al, pg. 6515).

On the epigenetic level, this modification could have occurred through inheritance or through a mutation in mitochondrial DNA, which lead to improper nuclear DNA replication or transcription. It could possibly be an inherited mitochondrial or nuclear DNA mutation as well. In order for any particular mutation or modification to become inheritable, the modified gene must be located in an individual's gamete cells (Carey, pg. 55). Also, it could have occurred through either DNA methylation or acetylation, or through a histone modification that later became methylated (Carey, pg. 73). The avenues are essentially endless as to how this particular deletion mutation occurred. The Todd et. al study did not specify the source of this mutation and so speculation is all that can be done here; until, of course, more research on the ADRA2b deletion mutation is conducted.

In the molecular level, mitochondria play a leading role. This is due to the fact that these organelles are the central hub of cortisol production (Picard et. al, pg. 76). Chronic stress and trauma cause cortisol to become dysregulated. When this occurs, a cascade of chemical reactions occur in the stress response systems of the brain, particularly in the hippocampus and its associated genes (Misfud and Reul, pg. 11337). The mitochondria are also responsible for the atrophy of the hippocampus in PTSD patients. This is due to genetic and non-genetic signalling of glucocorticoid receptors, in which modifications are made to the mitochondrial membrane (Feodorova and Sarafian, pg. 9). Cortisol is created by, and also modifies, the mitochondria, in ways that trickle down to the level of organs and organ systems.

Within the organ system level, modification to the ADRA2b gene is correlated to higher activity in the ventromedial prefrontal cortex, which in turn causes hyperactivity in the amygdala, and a reduction in the size of the hippocampus (Todd et. al, pg. 6514).

These same effects can be seen in these regions of the brain after the occurrence of trauma and are necessary for PTSD and other trauma spectrum disorders to occur (Anda et. al, pg. 182). This relates back to the dysregulation of cortisol that occurs during chronic stress. Secretion of cortisol is diminished due to changes in the HPA axis of the brain. Without cortisol, inflammation ensues, causing psychological disorders (Copeland et. al, pg. 7572).

The psychosocial and emotional level is characterized by hyperactivity in the amygdala, which is associated with intense mood and anxiety disorders, along with poor coping skills, and dysregulated cortisol secretion, which impacts behavioral responses to stressful stimuli. The Vaughn-Coaxum et. al study explains that the HPA axis, the prefrontal cortex, and the effect that trauma has on these regions of the brain, are the likely sources from which stress response and decision making are altered (Vaughn-Coaxum et. al, pg. 843). The study goes on to describe how the amygdala, in individuals with PTSD, has abnormal volume and thus affects an individual's processing of anxiety, fear, and mood. The prefrontal cortex itself is influenced by cortisol and other glucocorticoids that have a mediating effect on behavioral responses to sustained stress and trauma. Failure of the prefrontal cortex to remain plastic during stress was associated with failure to regulate emotions and behavior (Sinha et. al, pg. 8840). A lack of self-regulation is a hallmark of poor coping skills and poor mental health.

On the societal level, mental health issues are still a stigmatized topic of discussion due to a lack of understanding. Those with mental health disorders are more likely than the rest of the population to be repeated victims of violence because of the disenfranchisement that comes along with stigma. To reiterate the World Health Organization study: "Mental health issues are addressed in the report as both potential causes and consequences of these different types of violence" (Mercy et. al, pg. 21). The external displays of mental health disorders seem to elicit negative responses from others (Raine et. al, pg. 937). This occurs due to power plays of social stratification; an individual increases their social status by diminishing the already disenfranchised other (Copeland et. al, pg. 7572).

After an initial traumatic experience, an epigenetic tag alters the expression of a particular gene. The change in gene expression leads to dysfunctional production and secretion of molecules. The aberrant molecules then affect the functioning of organs and organ systems, such as the brain. When the molecular functioning of the brain is abnormal, the regulation of psychosocial and emotional states becomes abnormal as well. The individual, then, behaves abnormally, which elicits toxic responses from their society. These toxic responses lead to another traumatic experience and the cycle of revictimization begins again.

Discussion of Future Integrative Research

A biocultural view is paramount for all future research. Just because stressors can be reduced to the reactions of molecules does not mean that science is gaining a better understanding of how trauma works. As stated in the Graves article: "... the growing sophistication of modern biological techniques has not always allowed for the adoption of improved philosophical methodologies or ethical understanding regarding the applications of this new knowledge" (Graves, pg. 42). The body cannot be separated from the experiences of the individual within the body. Reducing human experience to molecules has the potential to disempower survivors' voices, to claim that biology, instead of personal perseverance, is the sole cause of their successes; to understand why revictimization occurs for some people and not for others, we cannot reduce everything to a genetic level because it takes away the autonomy and resilience of human beings to shape their own lives. Genetics can only do so much without the social environment to influence gene expression, behavior, health, biology, and sociocultural outcomes, as discussed by Harris and McDade in their synthesis of biological and sociological studies of human development (Harris and McDade, pg. 16). Simultaneously, we cannot blame the victims for their inability to prevent revictimization in their lives; the more exposure an individual has to trauma, the more likely negative outcomes will continue to occur (Vaughn-Coaxum et. al, pg. 854). In addition, factors such as structural inequality and biological damage are difficult to overcome, with or without traumatic experiences. This does not mean that some people are biologically determined to be trapped in cycles of trauma and revictimization; it means that different combinations of environmental and genetic factors have the potential to produce different results given the degree of attention that is known and paid to trauma and revictimization triggers. It is difficult to determine if an effective "treatment" for trauma and revictimization will ever be discovered, although there are promising avenues. There are, however, some issues with a "treatment" or "immunization" to stress from trauma. There is the potential for treatment of only the symptoms (Anda et. al, pg. 182) within individual bodies, while the focus is shifted away from treating adverse childhood experiences or society's relationship with violence (Lock, pg. 163). The tentative hypothesis of the biocultural trauma feedback loop, posited by this research, could be of use as a holistic avenue for future trauma research. A wide array of biological and sociological factors of trauma can be understood and addressed comprehensively, while keeping the lived experiences of survivors front and center. Utilizing this method, can lead to fuller, holistic, and more nuanced research on trauma and revictimization, as well as the myriad of avenues through which these occurrences are created and recreated.

Works Cited

Anda, R. F., Felitti, V. J., Bremner, J. D., Walker, J. D., Whitfield, C., Perry, B. D., . . . Giles, W. H. (2005). The enduring effects of abuse and related adverse experiences in childhood. European Archives of Psychiatry and Clinical Neuroscience, 256 (3), 174-186. doi:10.1007/s00406-005-0624-4

Becker-Blease, K. A., Turner, H. A., & Finkelhor, D. (2010). Disasters, Victimization, and Children's Mental Health. Child Development, 81 (4), 1040-1052.doi:10.1111/j.1467-8624.2010.01453.x

Carey, N. (2013). The epigenetics revolution: how modern biology is rewriting our understanding of genetics, disease, and inheritance. New York: Columbia University Press.

Cecil, C. A., Viding, E., Fearon, P., Glaser, D., & Mccrory, E. J. (2017). Disentangling the mental health impact of childhood abuse and neglect. Child Abuse & Neglect, 63 , 106-119. doi:10.1016/j.chiabu.2016.11.024

Chattarji, S., & Rao, R. P. (2014). Blood–brain biomarkers for stress susceptibility. Proceedings of the National Academy of Sciences, 111 (37), 13253-13254. doi:10.1073/pnas.1414663111

Copeland, W. E., Wolke, D., Lereya, S. T., Shanahan, L., Worthman, C., & Costello, E. J. (2014). Childhood bullying involvement predicts low-grade systemic inflammation into adulthood. Proceedings of the National Academy of Sciences, 111 (21), 7570-7575. doi:10.1073/pnas.1323641111

Elovainio, M., Pulkki-Råback, L., Hakulinen, C., Ferrie, J. E., Jokela, M., Hintsanen, M., . . . Keltikangas-Järvinen, L. (2015). Childhood and adolescence risk factors and development of depressive symptoms: The 32-year prospective Young Finns follow-up study. Journal of Epidemiology and Community Health, 69 (11), 1109-1117. doi:10.1136/jech-2014-205352

Feodorova, Y. N., & Sarafian, V. S. (2012). Psychological Stress – Cellular and Molecular Mechanisms. Folia Medica, 54 (3), 5-13. doi:10.2478/v10153-011-0091-9

Giberovitch, M., & Barry, R. (2014). Impact of Trauma: Vulnerability and Resilience. In Recovering from Genocidal Trauma (pp. 67-88). Toronto: University of Toronto.

Graves, J. L., Jr. (2015). Great Is Their Sin: Biological Determinism in the Age of Genomics. The Annals of the American Academy of Political and Social Science, 661 (1), 24-50. doi:10.1177/0002716215586558

Hagan, J., & Foster, H. (2001). Youth Violence and the End of Adolescence. American Sociological Review, 66 (6), 874. doi:10.2307/3088877

Harris, K. M., & Mcdade, T. W. (2018). The Biosocial Approach to Human Development, Behavior, and Health Across the Life Course. The Russell Sage Foundation Journal of the Social Sciences,4 (4), 2-26. doi:10.7758/rsf.2018.4.4.01

Howe, T. R. (2017). Families in Crisis: Violence, Abuse, and Neglect. In Marriages and Families in the 21st Century: A Bioecological Approach (2nd ed., pp. 357-402). Sage Publications.

Lock, M. (2015). Comprehending the Body in the Era of the Epigenome. Current Anthropology, 56 (2), 151-177. doi:10.1086/680350

Malabou, C. (2018). Is Psychic Phylogenesis Only a Phantasy? New Biological Developments in Trauma Inheritance. In Freud and Monotheism (pp. 177-198). Fordham University.

Mercy, J. A., Butchart, A., Dahlberg, L. L., Zwi, A. B., & Krug, E. G. (2003). Violence and Mental Health. International Journal of Mental Health, 32 (1), 20-35. doi:10.1080/00207411.2003.11449577

Mifsud, K. R., & Reul, J. M. (2016). Acute stress enhances heterodimerization and binding of corticosteroid receptors at glucocorticoid target genes in the hippocampus. Proceedings of the National Academy of Sciences, 113 (40), 11336-11341. doi:10.1073/pnas.1605246113

Moloney, R. D., Johnson, A. C., O'mahony, S. M., Dinan, T. G., Greenwood-Van Meerveld, B., & Cryan, J. F. (2016). Stress and the microbiota–gut–brain axis in visceral pain: relevance to irritable bowel syndrome. CNS neuroscience & therapeutics, 22(2), 102-117.

Motzkin, J. C., Philippi, C. L., Wolf, R. C., Baskaya, M. K., & Koenigs, M. (2015). Ventromedial Prefrontal Cortex Is Critical for the Regulation of Amygdala Activity in Humans. Biological Psychiatry, 77(3), 276-284. doi:10.1016/j.biopsych.2014.02.014

Picard, M., McManus, M. J., Gray, J. D., Nasca, C., Moffat, C., Kopinski, P. K., . . . Wallace, D. C. (2015). Mitochondrial functions modulate neuroendocrine, metabolic, inflammatory, and transcriptional responses to acute psychological stress. Proceedings of the National Academy of Sciences of the United States of America, 112 (48), 6614-6623. Retrieved March 5, 2019.

Picard, M., Mcewen, B. S., Epel, E. S., & Sandi, C. (2018). An energetic view of stress: Focus on mitochondria. Frontiers in Neuroendocrinology, 49 , 72-85. doi:10.1016/j.yfrne.2018.01.001

Pinti, M., Mussini, C., & Cossarizza, A. (2012). Mitochondrial DNA: A proinflammatory 'enemy from within' during HIV infection? Cell Death and Disease, 3 (5), 307. doi:10.1038/cddis.2012.47

Raine, A., Fung, A. L., & Lam, B. Y. (2011). Peer Victimization Partially Mediates the Schizotypy-Aggression Relationship in Children and Adolescents. Schizophrenia Bulletin, 37 (5), 937-945. doi:10.1093/schbul/sbr082

Reber, S. O., Siebler, P. H., Donner, N. C., Morton, J. T., Smith, D. G., Kopelman, J. M., . . . Halweg, A. L. (2016). Immunization with a heat-killed preparation of the environmental bacterium Mycobacterium vaccae promotes stress resilience in mice. Proceedings of the National Academy of Sciences of the United States of America, 113 (22), 3130-3139. Retrieved March 5, 2019.

Roberts, A. L., Gilman, S. E., Fitzmaurice, G., Decker, M. R., & Koenen, K. C. (2010). Witness of Intimate Partner Violence in Childhood and Perpetration of Intimate Partner Violence in Adulthood. Epidemiology, 21 (6), 809-818. doi:10.1097/ede.0b013e3181f39f03

Sas, K., Szabó, E., & Vécsei, L. (2018). Mitochondria, Oxidative Stress and the Kynurenine System, with a Focus on Ageing and Neuroprotection. Molecules, 23 (1), 191-218. doi:10.3390/molecules23010191

Shonkoff, J. P. (2017). Breakthrough Impacts: What Science Tells Us About Supporting Early Childhood Development. Young Children, 72 (2), 8-16. Retrieved March 5, 2019.

Sinha, R., Lacadie, C. M., Constable, R. T., & Seo, D. (2016). Dynamic neural activity during stress signals resilient coping. Proceedings of the National Academy of Sciences, 113 (31), 8837-8842. doi:10.1073/pnas.1600965113

Smith, D. (2019). The Epigenetics of Being Black and Feeling Blue: Understanding African American Vulnerability to Disease. In The Handbook of Research on Black Males (pp. 259-265). Michigan State University Press.

Todd, R. M., Ehlers, M. R., Muller, D. J., Robertson, A., Palombo, D. J., Freeman, N., . . . Anderson, A. K. (2015). Neurogenetic Variations in Norepinephrine Availability Enhance Perceptual Vividness. Journal of Neuroscience, 35 (16), 6506-6516. doi:10.1523/jneurosci.4489-14.2015

Turanovic, J. J., & Pratt, T. C. (2012). "Can't Stop, Won't Stop": Self-Control, Risky Lifestyles, and Repeat Victimization. Journal of Quantitative Criminology, 30 (1), 29-56. doi:10.1007/s10940-012-9188-4

Vaughn-Coaxum, R. A., Wang, Y., Kiely, J., Weisz, J. R., & Dunn, E. C. (2017). Associations Between Trauma Type, Timing, and Accumulation on Current Coping Behaviors in Adolescents: Results from a Large, Population-based Sample. Journal of Youth and Adolescence, 47 (4), 842-858. doi:10.1007/s10964-017-0693-5

Influence of Pinus Contorta and Arctostaphylos Uva-Ursi on the Presence of Cladonia Portentosa Subspecies Pacifica Within Protected Dune Ecosystems in Arcata, California

Courtney Michelle Harris (Humboldt State University), Kacia Layman (Humboldt State University), Jamie Wautlet (Humboldt State University), Grace Rhoades (Humboldt State University), Chris Belko (Humboldt State University)

Keywords: Pinus contorta, Arctostaphylos uva-ursi, Cladonia rangiferina, Cladonia portentosa subp. pacifica, presence, protected dune ecosystem, Arcata California., Ma-le'l Dunes, Lanphere Dunes unit, presence explained by elevation

Introduction

Lichens are generally known to be indicators of air quality and the presence of pollutants in an ecosystem. *Cladonia portentosa* subsp. pacifica, known commonly as reindeer lichen, or grey lichen, is a fungal composite symbiont that is composed of a mycobiont, or fungal partner, and a photobiont, in the form of a green algal partner [1]. *C. portentosa* subsp. pacifica is a light-colored fruticose lichen whose habitat range is limited to within several kilometers of the ocean and spans from Southeast Alaska to California. [2, 10]. The preferred substrate of *C. portentosa* subsp. pacifica includes sandy soils, humus, and moss over sand. In California, the habitat range of *C. portentosa* subsp. pacifica is limited to within a few kilometers of the ocean. *C. rangiferina* is another fruticose reindeer lichen in the same genus as *Cladonia portentosa* subsp. pacifica. These two lichens are macroscopically nearly identical to each other, excepting a few features which make them distinguishable. These features include different, subtle branching patterns, UV light reactivity in *C. portentosa* subsp. pacifica, and most notably their different habitat ranges. Although *C. portentosa* subsp. pacifica is limited to the coastal ranges, *C. rangiferina* grows in both hot and cold climates and thrives in areas of low disturbance, with low levels of moisture, as well as within areas that have well-drained, shallow soil and open canopies [3]. Both species are found in California, but *C. rangiferina* is most

Figure 1 - Lanphere Dunes survey site, top, and Ma-le'l Dunes survey site, bottom.

commonly prevalent in the arctic tundra regions of the northern hemisphere and in alpine regions and is believed to be a possible glacial refugia of the Pleistocene era [10]. Another documented difference between the two common taxa are the noted variability in the chemicals produced by them. *C. portentosa* subsp. pacifica, an exclusively coastal taxon, has been known to produce both usnic acid-containing and usnic acid-deficient chemotypes, whereas *C. rangiferina* has only been documented as an usnic acid-containing chemotype in its inland habitat range. This may suggest that selective pressures are less stringent in oceanic regions than in inland regions [10]. Due to the ability of *C. portentosa* subsp. pacifica and *C. rangiferina* to fix nitrogen, it would not be uncommon for both lichen species to develop in areas that experience fire as a natural regime, as this gives it the advantage of existing in habitats with low or non-existent levels of nitrogen [1]. *C. rangiferina* has a few commonly associated, fire-regime adapted, plant species that also share its ability to thrive in deficient soil conditions. Such areas that are occupied by the lichen, which also experience this natural fire regime, are predominantly lodgepole pine (*Pinus contorta*) forests with Kinnikinnick (*Arctostaphylos uva-ursi*) as a commonly associated species.

Although fire is infrequent in coastal regions, the shifting sand substrates and wind-borne salt spray act as a way to discourage competition and provide suitable conditions for stand formation of *Pinus contorta* [10]. Kinnikinnick is also tolerant of fire and is often indicative of sites that are moisture deficient due to rapid drainage [2, 4]. In the Pacific Northwest of Humboldt County, California, there are a few dense stands of pygmy lodgepole pines that exist in the swales of the coastal dunes ranging along the coastline within the unique dune habitat of the North Ma-le'l and Lanphere Dunes areas. Among these particular nutrient-deficient, well-draining, and sandy pygmy forests, kinnikinnick is also commonly present. Lodgepole pine and Kinnikinnick both exhibit fire, drought, and low-nutrient tolerance, and *Cladonia portentosa* subsp. pacifica tends to develop in undisturbed areas that have often experienced similar environmental coastal conditions [7]. The dune habitat where this lichen was found during this study appeared to be very limited to a small section of the coastal regions that harbored roughly the same necessary conditions for *C. rangiferina* to thrive. This suggests that there may be some aspect of specific areas of the coastal dune ecosystem that is allowing the subspecies of *C. portentosa* subsp. pacifica to exist and thrive, where its nearly identical relative, *C. rangiferina*, cannot.

The dune habitats observed in this study are currently wrapping up an extensive 15 year restoration process, and during that time the pygmy forest was left mostly undisturbed by human activity [5]. This restoration process was fueled by the desire to eradicate the invasive fore dune species of European beach grass, *Ammophila*, and the ice plant, *Carpobrotus*, using controlled burns on the dunes. These invasive plants work to trap most of the sand that blows off of the beach on the seaward slope rather than allowing the sand to flow naturally over the fore dune and into the semi-stable dunes behind it [5]. This effectively creates a protected area of habitat beyond the fore dune within the lower dune swales where other plant species, such as the ones observed in this study, have been established. It is the lack of disturbance within the pygmy forest from the invasive plants, the long term restoration projects currently underway, and the shared similar associated species of *C. ranigferina* that have led us to question what allows *C. portentosa* subsp. pacifica to reside in the dunes, but not its nearly identical relative. We hypothesized that if the presence of *Arctostaphylos uva-ursi* and *Pinus contorta* positively correlate with the presence of *C. portentosa* subsp. pacifica in areas near the ocean that have experienced relatively low disturbance, then the probability of *C. portentosa* subsp. pacifica being absent when *Arctostaphylos uva-ursi* and *Pinus contorta* are present in that range will be zero. Essentially, if we can find *P. contorta* and *A. uva-ursi* in a particular undisturbed oceanic area, then *C. portentosa* subsp. pacifica should also be present, if indeed the presence of the lichen subspecies is influenced by the presence of its nearly identical relative's commonly associated species.

Methods

To determine if there is a correlation between the three species of interest, we surveyed two large areas within a habitat that were known to contain the desired species (FIG 1). These sites were selected using inaturalist, an app that allows anyone who chooses to post a geo-tag of a specific species' location. We searched the app for areas where both the *Arctostaphylos* and the *Pinus* species were observed. We then surveyed twenty-two total sites at random within those two larger pre-determined areas of a similar habitat, and at each larger area we sampled eleven sites. The first large area was located in North Ma-le'l Dunes restoration habitat, and the second larger area was located North of Ma-le'l at the Lanphere Dunes Unit, both located in Arcata, California (FIG 2, 3).

At each location the sites were chosen at random by picking points sporadically on and off the trails, and each of us took turns choosing which direction to venture. The sites were predominantly nestled within the hind dunes, between the coastline, and Humboldt Bay, which were all located at lowland dune elevations within two kilometers of the ocean. At each site, the elevation, Longitude (N), and Latitude (W) was recorded using a GPS device, with precision and accuracy of 0.5-2 me-

Table 1. Candidate Models for Presence of Cladonia in Northern California

Candidate Models	AIC Value
Lichen~ elevation in feet + longitude + latitude + manzanita + lodgepole	31.63
Lichen~ elevation in feet + longitude + manzanita + lodgepole	29.69
Lichen~ elevation in feet + longitude + manzanita	27.96
Lichen~ elevation in feet	27.58
	27.31

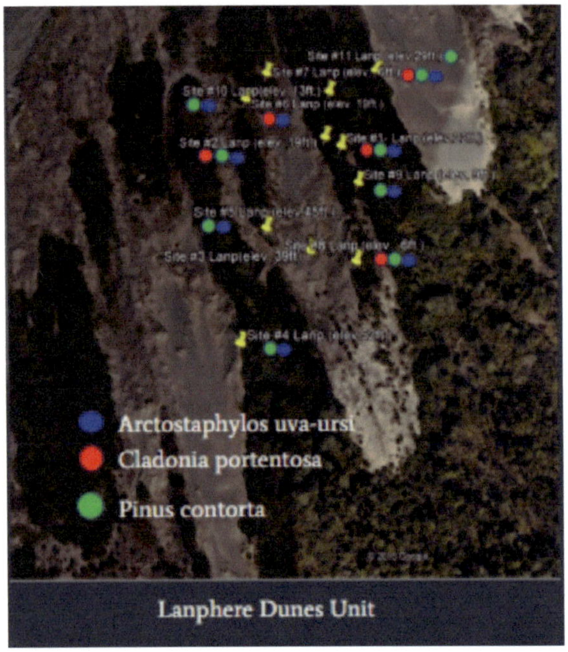

Figure 2 - Lanphere Dunes Survey Site. Yellow pins mark the location of sample taken, a blue dot represents presence of *A. uva-ursi,* a red dot represents presence of *C. portentosa subsp. pacifica,* and a green dot represents the presence of *P. contorta.* Each colored dot on the figure is representative of the species found at the site of the closest corresponding yellow pin.

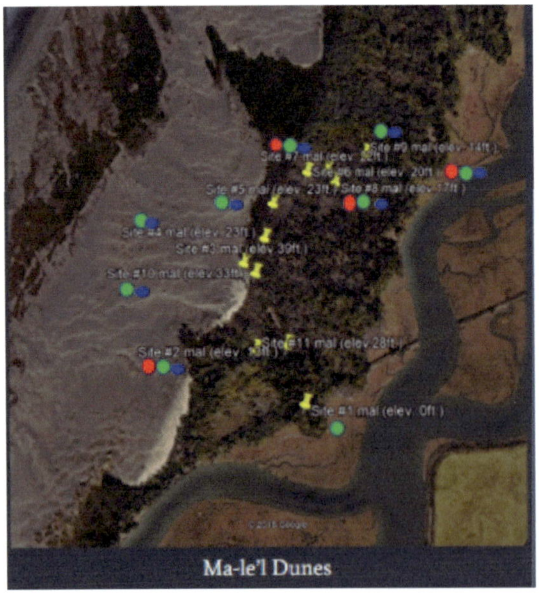

Figure 3 - Ma-le'l Dunes Survey Site. See caption for FIG 2 for key.

ters. Afterwards, the presence or absence of the desired species were determined by a visual survey of a 3-5 meter perimeter around the site. The data were analyzed using excel and r studio to perform a chi-square analysis and a stepwise regression of the candidate logistical models that we created to determine which variables had a significant effect.

Results

Pinus contorta and *Arctostaphylos uva-ursi* did not significantly influence presence of *Cladonia portentosa* subsp. pacifica, χ statistic= 24.84, p-value= 0.254, df=21. The presence of *Cladonia portentosa* subsp. pacifica was better explained by elevation, *p-value=0.0418, standard error=0.05021.* See Table 1 and Figure 4. Low Akike information criterion, or AIC values, show strongest correlation by using the data to estimate the quality of each model relative to the other models, see Table 1.

Discussion

Our data revealed that, although there was weak correlation between the presence of *A. uva-ursi* and *P. contorta,* and the presence of *C. portentosa* subsp. pacifica in areas where the data was collected, the correlation was not statistically significant. The patterns we observed after the chi-squared analysis led us to reject our hypothesis, and were more suggestive of a direct relationship between the presence of *C. portentosa* subsp. pacifica and elevation. This was based on the AIC values determined by a stepwise regression of all of the models we proposed in Table 1. This follows along with our research, which suggested that, within the dune ecosystems observed, the lichen might be less disturbed overall and thus more easily able to thrive in the lower-elevation protected swales of the dunes. Of the twenty-two sites that we surveyed, *C. portentosa* subsp. pacifica was more frequently present at relatively low elevations. Interestingly, our data concluded that for every one-foot increase of elevation the probability of the presence of *C. portentosa* subsp. pacifica decreased by 10%. Through our time-restricted research, we did not find any information regarding, specifically, how the lichen subspecies

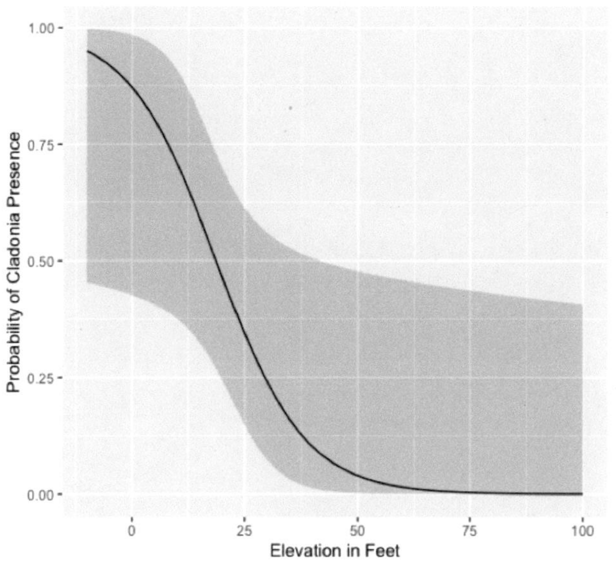

Figure 4 - Probability of *Cladonia portentosa subsp. pacifica* as Elevation (in ft.) increases. Shaded areas represent confidence intervals.

was introduced or naturalized into the dunes, if it is salt-tolerant, or what aspect of the dunes restricts its proliferation to some areas over others, even when environmental conditions and associated species remained constant. However, since our survey suggests that there may be a correlation between elevation and the presence of the lichen, this may hypothetically imply that, within the protected dune ecosystem that we observed, there is some specific factor(s) that may explicitly allow for this lichen subspecies to thrive. Ultimately, determining what dictates the presence or absence of *C. portentosa* subsp. pacifica in these surveyed areas may be worth investigating in future surveys to help scientists, stewards, and conservationists to better understand the sensitive dynamics of the dune ecosystems of Arcata, California. If we were to repeat this study, our future observations could greatly benefit from taking elevation and other associated factors, such as variation in chemotypes and the implications of that, into consideration when collecting and analyzing data.

Citations

[4] Crane, M. F. 1991. Arctostaphylos uva-ursi. In: Fire Effects Information System, [Online]. U.S. Department of Agriculture, Forest Service, Rocky Mountain Research Station, Fire Sciences Laboratory (Producer). Available: https://www.fs.fed.us/database/feis/plants/shrub/arcuva/all.html [2019, December 2].

[6] Coxson, D. S., & Marsh, J. 2001. Lichen chronosequences (postfire and postharvest) in lodgepole pine (Pinus contorta) forests of northern interior British Columbia. Canadian Journal of Botany, 79(12), 1449–1464. doi:10.1139/b01-127

[10] Goward, Trevor, and Teuvo Ahti. "Notes on the Distributional ecology of the Cladoniaceae (Lichenized Ascomycetes) in Temperate and Boreal Western North America." *J Hattori Bot. Lab.*, no. 82, July 1997, pp. 143–155, doi:10.18968/jhbl.82.0_143. Accessed Jan. 2020.

[7] Kershaw, K. A., & Rouse, W. R. 1971. Studies on lichen-dominated systems. II. The growth pattern of Cladonia alpestris and Cladonia rangiferina. Canadian Journal of Botany, 49(8), 1401–1410. doi:10.1139/b71-196

[2] Munger, Gregory T. 2008. Cladonia spp. In: Fire Effects Information System, [Online]. U.S. Department of Agriculture, Forest Service, Rocky Mountain Research Station, Fire Sciences Laboratory (Producer). Available: https://www.fs.fed.us /database/feis//lichens/claspp/all.html [2019, December 2].

[1] Nelson P.R., McCune B., Wheeler T., Geiser L.H., Crisafulli C.M. 2018 Lichen Community Development Along a Volcanic Disturbance Gradient at Mount St. Helens. In: Crisafulli C., Dale V. (eds) Ecological Responses at Mount St. Helens: Revisited 35 years after the 1980 Eruption. Springer, New York, NY

[3] Tegler, B., & Kershaw, K. A. 1980. Studies on lichen-dominated systems. XXIII. The control of seasonal rates of net photosynthesis by moisture, light, and temperature in Cladonia rangiferina. Canadian Journal of Botany, 58(17), 1851–1858. doi:10.1139/b80-214

[5] FINAL ENVIRONMENTAL ASSESSMENT. 2015. Humboldt Bay National Wildlife Refuge Sea-Level Rise Adaptation Demonstra-

tion Project Available: https://www.fws.gov/uploadedFiles/Region_8/NWRS/Zone_1/Humboldt_Bay_Complex/Humbolt_Bay/Sections/Documents/FinalHumboldtBayEAandSignedFONSI6-9-15.pdf

[8] iNaturalist. Available from https://www.inaturalist.org. Accessed November, 2019.

[9] "Alliance Detail Report: A3607" Usgs.gov .2013. Available from http://www1.usgs.gov/csas/nvcs/nvcsGetUnitDetails?elementGlobalId=899665

Analysis of the Variables Affecting Plant Species Richness in Deserts

Eli Kallison (Humboldt State University), Ellen Thompson (Humboldt State University), Maddison Keen (Humboldt State University), Rusty Newman (Humboldt State University)

Abstract

There are many hypotheses that attempt to explain patterns of species diversity in different environments. Deserts are a great place to study changes in species richness because they are relatively nutrient bare and exhibit low precipitation. This barebones environment means that slight shifts in climate and geography may lead to clear changes in species richness. We investigate how temperature, precipitation, water and light availability, latitude, elevation and other variables affect plant species richness in 20 deserts.

Keywords: biogeography, ecology, deserts, plants, botany, richness, R, multiple regression

Introduction

There are an abundance of hypotheses that attempt to explain factors that affect species richness across the globe (Channel and Lomolino 2000). These hypotheses vary drastically, and many are conflicting. For example, one purports that species richness increases in a habitat with a stable environment (Fordham et.al 2012) while another suggests that it increases in a stochastic environment (Yachi and Loreau 1999). Gaining a better understanding of what exactly affects plant species richness is important for conservation efforts as we face human-induced climate change. Our planet is continuing to experience an increasing extinction rate, and a population's probability of extinction is directly correlated with its variability in species richness (Channell and Lomolino 2000). We believe it is imperative to understand the variables that affect richness so that this knowledge can be used to preserve and increase richness.

Deserts are ideal for the study of species biodiversity patterns because richness varies greatly across desert regions (Vetaas and Grytnes 2002). They are also relatively nutrient bare and their climates are associated with high temperatures and limited precipitation year-round (Tucker et al. 2001). Because of these conditions, minor changes in climate variables can lead to clear shifts in species richness. Previous research suggests that temperature, precipitation, latitude, elevation, and sunlight are strongly implicated in affecting plant species richness. However, prior research seems unable to reliably isolate the magnitude of the impact that each variable has on richness. Additionally, because climatic variables are often correlated with one another, it is difficult to tease out the relative importance of each one. The objective of our study is to determine which climatic and geographical characters most significantly influence plant species richness in desert biomes. We predicted that water and light availability would have a significant effect on richness, as both variables are primary energy sources for plants. Our results, however, indicate that predicting plant species richness is explained by a much more complex model, explained in greater detail below.

Table 1. Variables that we collected and analyzed for each desert in the stepwise regression and description.

Variables Analyzed Per Desert	Description of Variable
HighT	Average High Temperature
LowT	Average Low Temperature
AvgT	Average Average Temperature
Months_Below_Freezing	Average Months in a Year Below Freezing
HighPrecip_inmon	Average High Precipitation (Inches/Month)
LowPrecip_inmon	Average Average Precipitation (Inches/Month)
AvgPrecip_inmon	Average Low Precipitation (Inches/Month)
HighDaylight_hrsday	Average High Daylight (Hours/Day)
LowDaylight_hrsday	Average Low Daylight (Hours/Day)
Latitude	Latitude
HighElevation	Highest Elevation
LowElevation	Lowest Elevation
ElevationDelta	Elevation Change
DesertSize	Size in Square Miles
PlantSpeciesQuantity	Number of Individual Plant Species

Methods

Data Collection

Research was conducted using publicly available online data from NOAA, national, state and regional park websites, and other available sources (Appendix). We researched 20 deserts across the globe, though many were in the United States (Appendix). We used the national park boundaries instead of the entire desert because plant inhabitants are better documented in the national parks. All plants present were counted towards our plant richness measure, including native, introduced, and invasive species. We measured nine variables (Table 1) that we hypothe-

Table 2. Desert size indices that were used for analysis

Desert Size (Square Miles)	Index Value
0-100	1
101-500	2
501-1,000	3
1,001-2,500	4
2501-10,000	5
10,001+	6

sized would most influence plant species richness in a desert ecosystem. Because desert size was so variable, we used indices to bucket groups of like-sized deserts together. This prevented the Desert Size variable from having an outsized effect on our model. For example, if the actual value of desert size was used for deserts such as the Gobi with 500,000 square miles of area, that variable would have an outsized effect on the model compared to smaller deserts, such as Carlsbad Cavern National Park with 73 square miles of area. The index allows us to consider desert size in our model but to not skew the results based on the big differences in area between our deserts. This was particularly important because our smaller deserts are the national parks that are often specifically protected because of their ecological importance, regardless of their small size. Therefore, it is not fair to compare 73 square miles of national park to 500,000 square miles of non-specific desert without an index.

Data Analysis: Stepwise Regression

To analyze the data we collected, we ran a stepwise regression analysis using the statistical program R (R Core Team 2013). Stepwise regression is a method of fitting regression models in which the choice of predictive variables is carried out by an automatic procedure. In each step, a variable is considered for addition to, or subtraction from, the set of explanatory variables known as a model. We compared our measured variables against plant species richness in our stepwise regressions, and Akaike information criterion (AIC) scores determined which combination of variables most significantly affects plant species richness. AIC is a measure of the quality of a model relative to each of the other models. It is used to compare models against one another by determining the relative amount of information lost by using one model instead of another. A model with a lower AIC value should be more efficient at producing the true probability distribution than a model with higher AIC value (Busemeyer and Diederich 2014). We used the model with the lowest AIC value to determine the best predictor of plant species richness. After we determined the best model, we ran linear regressions on each of the variables in the model.

Results

Table 3 shows our best-fit models along with the associated AIC values for each. This table shows the different combinations of variables that best explain plant species richness. Our best-fit model had an AIC value of 241.45 and included the following variables: average temperature, months below freezing, high precipitation, low precipitation, average precipitation, high daylight hours, latitude, high elevation and desert size.

Table 4 shows the results from the individual linear regressions for each variable from our best-fit model. This data explains the degree to which each variable is correlated to plant species richness (coefficient) and its significant (p-value) when viewed from outside the model. Variables that were positively correlated with plant species richness (in order from greatest to least) were: average precipitation, latitude, and high elevation. Variables that were negatively correlated with plant species richness (in order from least to greatest) were: low precipitation, high precipitation, months below freezing, and average temperature.

Table 3. AIC value model results from stepwise regression

Model	AIC
Model 1 (Plant Species Richness~AvgT + Months_Below_Freezing + HighPrecip_inmon + LowPrecip_inmon + AvgPrecip_inmon + HighDaylight_hrsday + Latitude + HighElevation + DesertSize)	241.45
Model 2 (Plant Species Richness~AvgT + Months_Below_Freezing + HighPrecip_inmon + LowPrecip_inmon + AvgPrecip_inmon + HighDaylight_hrsday + LowDaylight_hrsday + Latitude + HighElevation + DesertSize)	242.33
Model 3 (Plant Species Richness~HighT + AvgT + Months_Below_Freezing + HighPrecip_inmon + LowPrecip_inmon + AvgPrecip_inmon + HighDaylight_hrsday + LowDaylight_hrsday + Latitude + HighElevation + DesertSize)	243.66
Model 4(Plant Species Richness~HighT + LowT + AvgT + Months_Below_Freezing + HighPrecip_inmon + LowPrecip_inmon + AvgPrecip_inmon + HighDaylight_hrsday + LowDaylight_hrsday + Latitude + HighElevation + DesertSize)	245.26
Model 5(Plant Species Richness~HighT + LowT + AvgT + Months_Below_Freezing + HighPrecip_inmon + LowPrecip_inmon + AvgPrecip_inmon + HighDaylight_hrsday + LowDaylight_hrsday + Latitude + HighElevation + LowElevation + DesertSize)	246.41
Model 6(Plant Species Richness~HighT + LowT + AvgT + Months_Below_Freezing + HighPrecip_inmon + LowPrecip_inmon + AvgPrecip_inmon + HighDaylight_hrsday + LowDaylight_hrsday + Latitude + HighElevation + LowElevation + ElevationDelta + DesertSize)	246.41

Discussion

While our prediction that water and light availability would have the most significant effect on richness was correct in part, our analysis shows that a much more complex model is required to explain richness. Our best-fit Model 1 (Table 3) is the best predictor of plant species richness in the deserts that we studied and contains variables related to both light and water availability. Many of the variables included in this model exhibit intuitive results with regards to the magnitude and

Table 4. Linear regression results from top variables in our best-fit stepwise regression model with coefficients, standard error, test statistic, and p-values for each.

Variable From Stepwise Regression	Coefficient	Std. Error	t value	p-value
DesertSize	0.00	0.00	4.77	0.00
Latitude	28.88	6.74	4.28	0.00
Months_Below_Freezing	-167.60	44.04	-3.80	0.00
HighElevation	0.10	0.03	3.14	0.01
LowPrecip_inmon	-1380.00	667.40	-2.07	0.07
AvgT	-11.76	5.69	-2.07	0.07
AvgPrecip_inmon	1315.00	729.80	1.80	0.10
HighPrecip_inmon	-441.50	278.60	-1.59	0.14

direction of their effect on richness. For example, months below freezing and average temperature were negatively correlated with richness and average precipitation was positively correlated with richness. The negatively correlated variables behave as one might expect: the more months below freezing and/or the higher the temperature, the lower species richness in a desert. With more months below freezing, there is less water availability throughout the year. With higher temperatures in an already hot environment, conditions become inhospitable to life. Similarly, the higher the average precipitation in a desert is, the greater the species richness tends to be. Because deserts are water poor, having a higher baseline average precipitation probably allows for more favorable conditions for life.

In addition to these expected results, there are a few surprising findings. Most notably is that in our study deserts, having a higher average 'high precipitation' month every year and having a higher average 'low precipitation' month every year, are negatively correlated with plant species richness. This suggests that the greater our desert's lowest month of precipitation is—*and* the greater its highest month of precipitation is—the lower its species richness should be. In other words, the more precipitation there is in the desert's lowest and highest months of precipitation, the lower its richness is. Why would more rain during a desert's most extreme high and low precipitation months equate to lower richness? We hypothesize that with low amounts of precipitation and energy, the landscape would be dominated with smaller, less water-intensive plants. In these landscapes, there would be no big, dominant plants that consume most of the resources in the area (think redwood trees in a coastal forest). Without these dominant plants, a wider variety of species can compete for survival. This suggests that in conservation efforts aimed at maximizing species diversity, one should consider how undisturbed growth of dominant species might negatively affect overall richness.

Additionally, we found that although desert size did get incorporated into our model (and therefore was an important variable in predicting plant species richness), it did not present a coefficient when analyzed in isolation. This would suggest that desert size may not be a good predictor of plant species richness on its own. This is surprising because, intuitively, more area means more room for a greater diversity of plants to potentially grow. Instead, desert size is a useful predictor only in conjunction with all the other environmental variables incorporated in a model. In biogeographic literature pertaining to species conservation, there are two fundamental approaches to consider in terms of area of habitat: single large or several small plots (Järvinen, 1982). Our research seems to support the idea, as do many other studies, that smaller noncontiguous habitats are just as good at retaining high species richness numbers as larger contiguous habitats. To confirm findings from our research, we would like to see similar analyses performed in different biomes to see how the variables affected plant species richness differ in different environments.

In conclusion, predicting plant species richness in deserts is complex and is influenced by a host of potentially interconnected variables. The possibility of variables being influenced by each other makes definitive analysis in this complex environment even more difficult. In each of our deserts, though, we see the same variables playing similar crucial roles in determining species richness and overall production of the ecosystem.

While deserts are a wonderful environment that showcase an incredibly diverse array of adaptations, they are also harsh climates that are difficult to survive in. Currently, a phenomenon called desertification, or the spread of deserts to regions that were not previously desert, is a growing threat (Darkoh 1998). As such, it is important to better understand deserts and the variables that govern and influence plant life within them. In addition to illuminating factors affecting biodiversity in general, our research also provides valuable information that can help us understand the variables that govern the deserts in our study. In turn, this information can help us to better prepare for hotter environments and desertification caused by increasingly rapid climate change. With a better understanding of these crucial factors, we as humans can better protect and restore the planet and allow life to flourish.

References

Busemeyer, R. J, Diederich A. 2014. Chapter 4 - Estimation and Testing of Computational Psychological Models. Neuroeconomics:49-61.

Channell, R., and M. V. Lomolino. 2000. Dynamic biogeography and conservation of endangered species. Nature 403:84–86.

Darkoh, M. B. K. 1998. The nature, causes and consequences of desertification in the drylands of Africa. Land Degradation and Development 9:1 1-20.

Fordham, D. A., Resit Akçakaya, H. , Araújo, M. B., Elith, J. , Keith, D. A., Pearson, R. , Auld, T. D., Mellin, C. , Morgan, J. W., Regan, T. J., Tozer, M. , Watts, M. J., White, M. , Wintle, B. A., Yates, C. and Brook, B. W. 2012. Plant extinction risk under climate change: are forecast range shifts alone a good indicator of species vulnerability to global warming? Glob Change Biol, 18: 1357-1371.

Järvinen, Olli. "Conservation of endangered plant populations: single large or several small reserves?." Oikos (1982): 301-307.

Jonathan P. Comstock, J. R. E. 1992. Plant adaptation in the Great Basin and Colorado Plateau. The Great Basin Naturalist 52:195–215.

Martin L. C. 1989. Growth-form diversity and community structure in desert plants. J. Arid Environ 17:199-209.

Tucker, C. J., D. A. Slayback, J. E. Pinzon, S. O. Los, R. B. Myneni, and M. G. Taylor. 2001. Higher northern latitude normalized difference vegetation index and growing season trends from 1982 to 1999. International journal of biometeorology 45:184–190.

Vetaas, O. R., and J.-A. Grytnes. 2002. Distribution of vascular plant species richness and endemic richness along the Himalayan elevation gradient in Nepal. Global Ecology and Biogeography 11:291–301.

Yachi S, Loreau M. 1999. Biodiversity and ecosystem productivity in a fluctuating environment: The insurance hypothesis. Proceedings of the National Academy of Sciences. 96:1463-1468.

Appendix

Table 1A. List of deserts used in study.

Desert Name
Carlsbad Caverns National Park
Nambung National Park
Arches National Park
Great Basin National Park
Saguaro National Park
Petrified Forest National Park
Zion National Park
Capitol Reef National Park
Little Desert National Park
Organ Pipe Cactus NM
Canyonlands National Park
Joshua Tree National Park
Big Bend National Park
Grand Canyon National Park
Kobuk Valley National Park
Death Valley National Park
Atacama
Thar
Kalahari
Gobi

Steps to Collecting Data

1. Determine latitude via google.
2. Get temperature and precipitation data from NOAA, google aggregated data.
    a. Look at the months, pick out the lowest of the low temp value
    b. Look at all the months, pick out the highest of the high temp value
3. Get precipitation via US climate data, also aggregated by google.
    a. Look at the months, pick out the lowest
    b. Look at all the months, pick out the highest
    c. Here is a good source for precipitation data: https://www.timeanddate.com/weather/@5536441/climate
4. Google monthly high precipitation, or high annual divided by 12.
5. Google monthly low precipitation, or low annual divided by 12.
6. Google monthly average temperature, or average annual divided by 12.
7. Get daylight hours data
    a. https://www.timeanddate.com/sun/

    b. Slide the line on the graph and determine the high, low, and average *Daylight Total* for the year 2018, this is shown below the graph in the middle.
8. Look on the National Park Service website for elevation
    a. If you don't find it there, record on the *Sources* google doc where you found the elevation data
9. Look under the *Plants* page on the National Park Service website for the number of plant species

Data Sources

Best time to go to, climate and weather Kobuk Valley National Park, Alaska. (2018). https://www.best-travelmonths.com/united-states/alaska/kobuk-valley-national-park-4030757/.
canyonlands-national-park-map.jpg (3369×4537). (2018). http://npmaps.com/wp-content/uploads/canyonlands-national-park-map.jpg.
capitol-reef-national-park-map.jpg (2861×3899). (2018). http://npmaps.com/wp-content/uploads/capitol-reef-national-park-map.jpg.
Climate & Weather Averages in Canyonlands National Park, Utah, USA. (2018). https://www.timeanddate.com/weather/@5536441/climate.
Climate & Weather Averages in Capitol Reef National Park, Utah, USA. (2018). https://www.timeanddate.com/weather/@5536448/climate.
Climate & Weather Averages in Little Desert, Victoria, Australia. (2018). https://www.timeanddate.com/weather/@2159999/climate.
Climate & Weather Averages in Petrified Forest National Park, Arizona, USA. (2018). https://www.timeanddate.com/weather/@5308614/climate.
Climate & Weather Averages in Saguaro National Park, Arizona, USA. (2018). https://www.timeanddate.com/weather/@5312529/climate.
Flora & Fauna | Death Valley | Oh, Ranger! (2018). http://www.ohranger.com/death-valley/flora-fauna.
Kalahari Desert Plants - Kalahari Desert. (2018). http://www.kalaharidesert.net/Kalahari-Desert-Plants.html.
NPSpecies- Search for a Park Species List. (2018). https://irma.nps.gov/NPSpecies/Search/SpeciesList/PEFO.
NPSpecies- Search for a Park Species List. (2018). https://irma.nps.gov/NPSpecies/Search/SpeciesList/CARE.
phone, please email us at zion_park_information@nps gov C. U. (2018). Plants - Zion National Park (U.S. National Park Service). https://www.nps.gov/zion/learn/nature/plants.htm.
ProTrails | Golden Throne, Capitol Gorge Trailhead, Capitol Reef National Park, Utah. (2018).

https://www.protrails.com/trail/828/capitol-reef-national-park-golden-throne.
Sunrise and sunset times in Canyonlands National Park. (2018). https://www.timeanddate.com/sun/@5536441.
Sunrise and sunset times in Capitol Reef National Park. (2018). https://www.timeanddate.com/sun/@5536448.
Sunrise and sunset times in Little Desert National Park. (2018). https://www.timeanddate.com/sun/@7910192.
Sunrise and sunset times in Petrified Forest National Park. (2018). https://www.timeanddate.com/sun/@5308614.
Sunrise and sunset times in Saguaro National Park. (2018). https://www.timeanddate.com/sun/@5312529.
These Striking Rock Formations Are up to 270 Million Years Old. 2009, November 5. https://www.nationalgeographic.com/travel/national-parks/capitol-reef-national-park/.
Thomas, K. A. (2018). Part I: Vegetation of Petrified Forest National Park, Arizona:126.
Tucson, M. A. 3693 S. O. S. T., and A. 85730 P.-5153 C. Us. (2018). Nature - Saguaro National Park (U.S. National Park Service). https://www.nps.gov/sagu/learn/nature/index.htm.
Victoria, National Parks Service, Victoria, and Department of Natural Resources & Environment. 1996. Little Desert National Park management plan. The Department, East Melbourne.
Weather and climate for Zion National Park, Utah. (2018). http://www.americansouthwest.net/utah/zion/weather.html.

Exploration of Antiviral Effects of RNA-dependent RNA Polymerases 3, 4 and 5 in *Arabidopsis*

Eli Kallison (Humboldt State University), Aleksandra Berici (University of Missouri, Coulmbia), Blake C. Meyers (University of Missouri, Coulmbia)

Abstract

Small RNAs play a number of vital roles in plants, including viral resistance. A particular class of small RNA that regulates plant defense from pathogens, among other functions, is known as small interfering RNAs (siRNAs). A key player in the biogenesis pathway of siRNAs are RNA-Dependent RNA Polymerases (RDRs). There are six RDR proteins in Arabidopsis thaliana, three of which have well documented roles (RDRs 1, 2 and 6), and three of which have no documented role. Here, we look at the functions of the unstudied RDRs, RDRs 3, 4 and 5, to see whether they affect antiviral resistance.

Keywords: small RNA, siRNA, RNA dependent RNA polymerases, plant biology, molecular biology, molecular genetics, plant pathology

Introduction

The biggest challenge faced in modern agriculture is increasing food production to meet the demands of a growing population. Disease plays a major role in crop failure, with losses to staple crops due to pests and disease ranging between 10-40% [1], [2]. This effect is being compounded by plants' increased susceptibility to disease in the face of global climate change [3]. As such, understanding the factors that increase plant resistance to disease is crucial to our efforts towards sustainable agriculture. One of the key viral resistance mechanisms in plants is RNA-mediated silencing. This mechanism is carried out by small RNAs (sRNA), noncoding signaling molecules that range between 21 and 24 nucleotides in length. Small RNAs play essential roles in plant development, reproduction and defense [4]. In particular, those that regulate defense responses in plants belong to a class known as small interfering RNAs (siRNAs) [5].

These siRNAs are produced from a double-stranded RNA (dsRNA) precursor, which is cleaved by a DICER-LIKE (DCL) protein. siRNAs then form a complex with ARGONAUTE

(AGO) proteins and target complementary nucleotide sequences, effectively silencing the expression of the target gene [6]. A major step in siRNA biogenesis, the precursor synthesis, is what differentiates siRNAs from other sRNAs. The dsRNA precursor is a product of RNA-dependent RNA polymerase (RDR), which uses single-stranded RNA (ssRNA) as a template to generate the second strand [5].

RDR is an ancient protein family that is intrinsically linked with the evolution of RNA viruses and plays essential roles in their RNA replication [7]. RDRs have been conserved throughout the evolution of plants, functioning to synthesize dsRNA from ssRNA [6]. There are 6 different types of RDRs found in *Arabidopsis thaliana* [8]. RDRs 1, 2 and 6 are well studied, and have been shown to generate siRNAs involved in mediating plant stress response, pathogen resistance, female gamete formation, transgene silencing, and plant development [9]. However, the roles of RDRs 3, 4 and 5 and their sRNA products have not been well explored to date [8].

Being so highly conserved across kingdoms, and because of the vital roles that RDRs 1, 2 and 6 play, we believe that studying the functions of RDRs 3, 4 and 5 is essential. Two major lines of evidence indicate that RDR-derived sRNA silencing forms the core of an antiviral defense in plants: (1) siRNAs targeting viral RNA accumulate during infections; and (2) viruses produce virulence factors called viral suppressors of RNA silencing to counteract this defense [10]. This raises a question as to what role, if any, do RDRs 3, 4 and 5 play in antiviral defense.

Materials and Methods

Plant Materials

We used nine different *Arabidopsis thaliana* genotypes, including wildtype *Columbia-0 (Col-0)* and the following mutants: *rdr1-1*, *rdr2-1*, *rdr3-2*, *rdr4-2*, *rdr5-3*, *rdr6-4*, *ago2-1*, and *dcl2/3/4* [11]. All the mutant lines were received from the Carrington lab at the Donald Danforth Plant Science Center and are in *Col-0* background. Plants were grown in growth chambers under long day conditions (16-hour light/8-hour dark) at 21°C and 50% relative humidity. The *rdr3, 4* and *5* mutants used in this study are SALK lines (Accession numbers: Salk 036925, Salk 088175, Salk 023522).

Nicotiana benthamiana plants were used to establish viral inoculum used for *Arabidopsis* infection. *Nicotiana* plants were grown in greenhouses at 28°C (day temperature) and 25°C (night temperature) at 60% relative humidity. Watering was stopped two days pre infiltration to increase leaf permeability.

DNA Plasmids

DNA plasmids containing CRISPR constructs were generated to create the following knockout lines: *rdr3, rdr4, rdr5, rdr3/4, rdr3/5, rdr4/5 and rdr3/4/5.* Guide RNA constructs were assembled using the Goldengate method. The constructs were inserted into the destination vector containing a Cas9 coding sequence via Gateway LR reaction. The inserts were confirmed by sequencing and final plasmids were transformed into *Agrobacterium tumefaciens* LBA4404.

TCV Virus Infection Assays

Plasmids carrying TCV, TCV-CPB and P19 were shared by Dr. Zheng [12]. Inoculum preparation and inoculation of *Arabidopsis thaliana* were performed as described [12]. Briefly, *Nicotiana benthamiana* leaves were infiltrated with each of the infection clones. Infected leaves were collected 5 days after infiltration (DAI) and ground in 200mM NaOAc. After overnight incubation at 4°C in a PEG-8000 and NaCl solution, virions were pelleted by centrifugation and resuspended in 10mM NaOAc. This inoculum stock was diluted 10 times with 10mM NaOAc and used to inoculate the 4 largest *Arabidopsis* rosette leaves.

TuMV Virus Infection Assay

Plasmids carrying TuMV-GFP were shared by the Carrington lab [13]. *Nicotiana benthamiana* plants were infiltrated as described previously for inoculum preparation [14]. Infection development was confirmed by GFP fluorescence under UV light at 6DAI. Infected leaves were collected and inoculum prepared as described [13].

Image Capture

Images for phenotyping were captured by Raspberry Pi cameras mounted directly over the plants in the growth chamber. Plants of different treatment and genotypes were placed randomly throughout the chamber and assigned a code for identification purposes. The cameras were programmed to capture one image every hour starting from 2 days pre-infection to 14 days post-infection. One camera was used per tray. Black mesh was placed around the plants to eliminate background interference.

Image Analysis

Images taken every day at 11am were analyzed using PlantCV, a python-based phenotyping program. PlantCV was used to identify and segment plant leaves and distinguish them from the background mesh. Spots of necrosis (brown) and chlorosis (yellowing) were identified against healthy leaf tissue (green). Pixel counts for healthy and unhealthy tissue were collected for each plant.

Statistical Analysis

All statistical analysis was performed in R. Proportion of necrotic tissue per day was calculated for each plant. Averages of necrotic tissue proportion were taken per genotype for each treatment each day. Kolmogorov-Smirnov tests were run for each genotype compared to Col-0 wt to test whether the distribution curves were significantly different from one another.

Results

Image-Based Analysis of Disease Symptoms

Plants that were infected by TuMV and TCV showed more severe symptoms than those infected with TCV-CPB. Overall, we found no evidence to suggest that RDRs 3, 4 or 5 play any role in antiviral defense in *Arabidopsis*.

When graphed, our positive control mutant, *dcl2/3/4*, showed the greatest necrotic tissue expansion across all viral treatments, while our positive control, wt Col0, showed the least necrotic tissue expansion across almost all treatments (see Figure 1).

We used the Kolmogorov-Smirnov (K-S) test to determine whether or not the distributions for each genotype were different than our Col0 control. The output from the K-S test is a p-value that states whether or not two series of data were drawn from the same distribution. For most of our genotypes across treatments, p-values did not meet our required significance level of 0.05 (see Table 1). However, we did see that *rdr5* mutants had a significantly different distribution curve than Col0 after TCV infection (p-value: .0015). We also saw that *rdr4* mutants had significantly different distributions than Col0 after Mock and TCV infections, though necrotic tissue progression continued more slowly than Col0 in the case of TCV.

RDR4 Mutants Exhibit Significant Developmental Delay

rdr4 mutants consistently displayed stunted growth when compared to the other mutants that we observed. Plants were significantly smaller than wild-type Col0 of the same age (p-value: 3.57×10^{-7}, see Figure 2).

Discussion

Although we completed only one full replicate, we found no significant evidence that RDRs 3, 4 and 5 function in viral defense in plants. We think that the significant p-values for *rdr4* mutants in Table 1 are due to inconsistencies in their growth rate and that their stunted growth doesn't allow for accurate analysis of proportion of necrotic tissue in the plant. We attribute this to there being insufficient total tissue area for the calculation to be meaningful without more replicates. Although *rdr5* mutants exhibited a significant p-value compared to Col0 after TCV infection, the plants started out with more

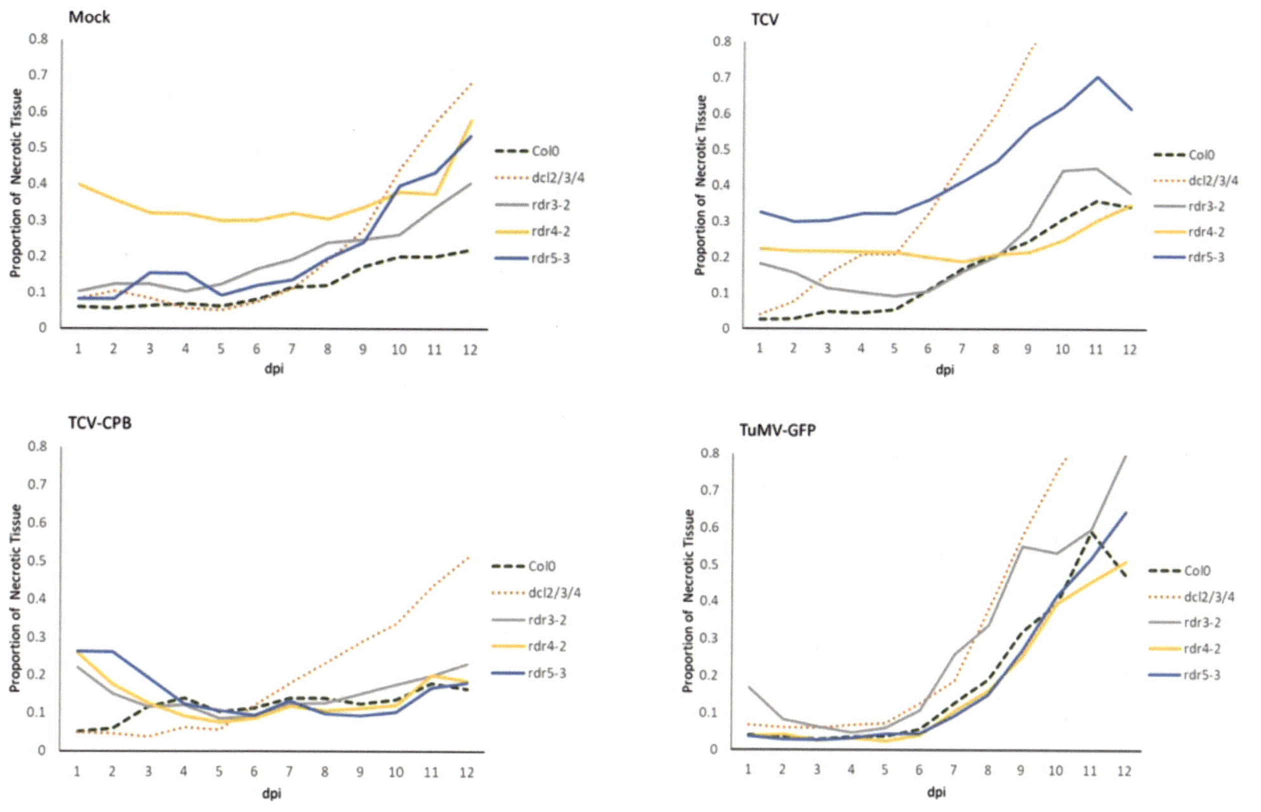

Figure 1. Line graphs showing the progression of necrotic tissue expansion over time after infection of *Arabidopsis* with Mock, TCV, TCV-CPB, and TuMV-GFP viruses.

Table 1. P-values from K-S test for each genotype against Col0 wildtype, separated by treatment.

Genotype	Mock p-values	TCV p-values	TuMV p-values	TCV-CPB p-values
dcl2/3/4	0.54	0.10	0.10	0.10
rdr3	0.10	0.26	0.26	0.54
rdr4	7.40E-07	0.03	1.00	0.87
rdr5	0.10	1.50E-03	1.00	0.54

necrotic tissue than Col0 and increased at almost the exact same rate as infection progressed. This significant difference, then, is not due to rate of necrotic tissue accumulation.

Because the virus-infected single mutant rdr knockout lines showed no sign of increased susceptibility to viruses, we can conclude that there must be alternative pathways for viral resistance in plants. With evidence to support alternative viral resistance pathways, we propose three likely hypotheses: (1) RDRs function redundantly and other RDRs are able to compensate for the lack of RDR3, 4 or 5; (2) antimicrobial compounds, secondary metabolites and other chemical compounds sufficiently defend plants from viruses; and (3) RDRs 3, 4 and 5 don't function in antiviral defense but do have other functions.

Hypothesis 1: RDRs function redundantly.

Figure 2. Comparison of 2 week old Col0 wildtype, the day before infection, grown in the same conditions.

There is evidence to support the idea that RDRs function redundantly. RDR1 and RDR6 have both been shown to play a role in antiviral defense [9]. Further, RDRs 3, 4 and 5 are tandem repeats in the *Arabidopsis* genome, suggesting that they arose from a duplication event somewhere along their evolutionary pathway [15]. Because of this potential duplication event, it stands to reason that they might function in the same way. Sequencing the transcriptome of single mutant knockouts could show whether other RDR transcripts were upregulated in the absence of the single RDR that was knocked out. If the RDRs do indeed function redundantly, then it would require multiple knockout mutants (double, triple and even hextuple) to see any effect. We plan to create these mutants and repeat the experiments with the multiple knockout lines.

Hypothesis 2: Antimicrobial compounds, secondary metabolites, and other chemical compounds sufficiently defend plants from viruses. We know that secondary metabolites and antimicrobial compounds protect plants from herbivores and other dangers, so it is likely that they also play a role in viral defense [16].

Hypothesis 3: RDRs 3, 4 and 5 don't function in antiviral defense but do have other functions. From our initial testing, *rdr4* mutants seem to show signs of stunted growth and development. It has also been found that RDR 6 plays a role in eukaryotic cell development, so this would be a natural line of inquiry [9]. It is also possible that RDRs 3, 4 and 5 each have one or many additional roles that we are currently unaware of.

Our most striking result was the discrepancy in growth rate and size between *rdr4* mutant and wildtype Col0. As a result of these findings, we suspect that RDR4 plays a role in growth and development of *Arabidopsis* because of the statistically significant size differences. On average, the *rdr4* mutant was three and a half times smaller than wt at two weeks old before viral infection. We hypothesize that there are transcripts that regulate growth that are targeted by RDR4's siRNA product. Exploring this, possibly by generating and comparing RNA-seq data between the various RDR mutants, seems like a fruitful avenue for further research. Continued research relating to RDR4's role in growth and development, and on the antiviral effects of RDRs 3, 4 and 5, will continue to be important in agriculture and horticulture. The more we understand about disease resistance during the growth and development of plants, the more efficiently we can engineer our agricultural systems to feed our growing population. Beyond the practical benefits to agriculture, research on the RDRs in *Arabidopsis* will provide further insight into the fundamental molecular processes of their small RNA products. This insight can be leveraged by other scientists to enable deeper research that may provide additional scientific breakthroughs and practical applications.

References

[1] S. Savary, A. Ficke, J.-N. Aubertot, and C. Hollier, "Crop losses due to diseases and their implications for global food production losses and food security," *Food Secur.*, vol. 4, no. 4, pp. 519–537, Dec. 2012, doi: 10.1007/s12571-012-0200-5.

[2] S. Savary, L. Willocquet, S. J. Pethybridge, P. Esker, N. McRoberts, and A. Nelson, "The global burden of pathogens and pests on major food crops," *Nat. Ecol. Evol.*, vol. 3, no. 3, p. 430, Mar. 2019, doi: 10.1038/s41559-018-0793-y.

[3] P. K. Anderson, A. A. Cunningham, N. G. Patel, F. J. Morales, P. R. Epstein, and P. Daszak, "Emerging infectious diseases of plants: pathogen pollution, climate change and agrotechnology drivers," *Trends Ecol. Evol.*, vol. 19, no. 10, pp. 535–544, Oct. 2004, doi: 10.1016/j.tree.2004.07.021.

[4] A. F. A. Samad *et al.*, "MicroRNA and Transcription Factor: Key Players in Plant Regulatory Network," *Front. Plant Sci.*, vol. 8, 2017, doi: 10.3389/fpls.2017.00565.

[5] D. Baulcombe, "RNA silencing in plants | Nature," 2004. [Online]. Available: https://www.nature.com/articles/nature02874. [Accessed: 11-Jun-2019].

[6] R. W. Carthew and E. J. Sontheimer, "Origins and Mechanisms of miRNAs and siRNAs," *Cell*, vol. 136, no. 4, pp. 642–655, Feb. 2009, doi: 10.1016/j.cell.2009.01.035.

[7] S. T. de Farias, A. P. dos Santos Junior, T. G. Rêgo, and M. V. José, "Origin and Evolution of RNA-Dependent RNA Polymerase," *Front. Genet.*, vol. 8, Sep. 2017, doi: 10.3389/fgene.2017.00125.

[8] Q. Fei, R. Xia, and B. C. Meyers, "Phased, Secondary, Small Interfering RNAs in Post-transcriptional Regulatory Networks," *Plant Cell*, vol. 25, no. 7, pp. 2400–2415, Jul. 2013, doi: 10.1105/tpc.113.114652.

[9] M. R. Willmann, M. W. Endres, R. T. Cook, and B. D. Gregory, "The Functions of RNA-Dependent RNA Polymerases in Arabidopsis," *Arab. Book*, vol. 9, p. e0146, Jan. 2011, doi: 10.1199/tab.0146.

[10] S.-W. Ding and O. Voinnet, "Antiviral Immunity Directed by Small RNAs," *Cell*, vol. 130, no. 3, pp. 413–426, Aug. 2007, doi: 10.1016/j.cell.2007.07.039.

[11] E. Allen, Z. Xie, A. M. Gustafson, and J. C. Carrington, "microRNA-Directed Phasing during Trans-Acting siRNA Biogenesis in Plants," *Cell*, vol. 121, no. 2, pp. 207–221, Apr. 2005, doi: 10.1016/j.cell.2005.04.004.

[12] X. Zheng, N. Fahlgren, A. Abbasi, J. C. Berry, and J. C. Carrington, "Antiviral ARGONAUTEs Against Turnip Crinkle Virus Revealed by Image-based Trait Analysis," *Plant Physiol.*, p. pp.00121.2019, May 2019, doi: 10.1104/pp.19.00121.

[13] H. Garcia-Ruiz *et al.*, "Arabidopsis RNA-Dependent RNA Polymerases and Dicer-Like Proteins in Antiviral Defense and Small Interfering RNA Biogenesis during Turnip Mosaic Virus Infection[W][OA]," *Plant Cell*, vol. 22, no. 2, pp. 481–496, Feb. 2010, doi: 10.1105/tpc.109.073056.

[14] X. Li, "Infiltration of Nicotiana benthamiana Protocol for Transient Expression via Agrobacterium," *BIO-Protoc.*, vol. 1, no. 14, 2011, doi: 10.21769/BioProtoc.95.

[15] J. Zong, X. Yao, J. Yin, D. Zhang, and H. Ma, "Evolution of the RNA-dependent RNA polymerase (RdRP) genes: Duplications and possible losses before and after the divergence of major eukaryotic groups," *Gene*, vol. 447, no. 1, pp. 29–39, Nov. 2009, doi: 10.1016/j.gene.2009.07.004.

[16] A. R. War *et al.*, "Mechanisms of plant defense against insect herbivores," *Plant Signal. Behav.*, vol. 7, no. 10, pp. 1306–1320, Oct. 2012, doi: 10.4161/psb.21663.

Differences in Algal, Gastropod, and Arthropod Coverage Surrounding Pisaster Ochraceus and Other Asteroids in Humboldt County, California

Desa Smith (Humboldt State University), Jana Lott (Humboldt State University), Savannah Williams (Humboldt State University), Shayan Farzadpour (Humboldt State University)

Abstract

Predation within the intertidal zone, specifically regarding the predator-prey interaction between sea stars in Class Asteroidea and their food source, contributes to a large biodiversity of organisms. In this experiment, we compared the percent coverage of algae and shelled organisms within a quadrat surrounding three types of predatory sea stars (Pisaster ochraceus, Dermasterias imbricata, and Lepasterias hexactis) to determine their habitat based on predatory behavior. Our data showed a significant difference in algae and shelled organism coverage between different species of sea star; shelled organism percent coverage was less than algae percent coverage in all species of sea stars. The niche, or role and position a species plays in its environment, of each star ultimately influenced the frequency and diversity of other recorded organisms.

Keywords: asteroids, algae coverage, gastropod, arthropod, intertidal, Pisaster ochraceus

Introduction

The beaches in Northern California consist of rocky intertidal shores, which are inhabited by various marine invertebrates and a wide selection of algae (Foster et al. 1988). The intertidal zone can be broken up into four smaller zones, described by the species that inhabit those zones (Foster et al. 1988). The two lowest zones, the Splash Zone and the Barnacle Zone, are mostly comprised of kelp, sea stars, mussels, and barnacles. One of the most common inhabitants in Northern California is the ochre star *Pisaster ochraceus*, accompanied by the leather star (*Dermasterias imbricata*) and six-rayed star (*Lepasterias hexactis*). Asteroids are carnivorous, and their diet consists mostly of shelled organisms, such as mussels (Paine 1976, Penchaszadeh and Lera 1983, Monteiro and Pardo 1994). These sea stars are often found near areas of dense mussel beds or other food sources (Paine 1976).

Table 1. Summary of F test comparing the variance between the different groups (F = f distribution factor, df = degrees of freedom).

Comparison Groups	F	Num df	Denom df	P-value
Algae: Pisaster v Other	1.2661	49	37	0.458
Shelled: Pisaster v Other	5.1344	33	22	0.0001675
Pisaster: Algae v Shelled	284.7	49	33	< 2.2e-16
Other: Algae v Shelled	1154.6	37	22	< 2.2e-16

Sea stars play an important role in the intertidal zone, they are described as a keystone species based on how their presence increases biodiversity due to their predation on mussels (Manzur et al 2009). Biodiversity is crucial to keeping an ecosystem functioning, as each species has their own role in the productivity of that ecosystem. In this study, we looked at different species of sea stars and the habitats immediately surrounding them by measuring algae coverage and animal coverage in that area. Animals that were observed in this area included gastropods such as marine snails and limpets, and arthropods such as hermit crabs. With the impending threat of climate change, intertidal species are at a greater risk of desiccation, or drying out, which can threaten the species survival. Tracking sea star's prey selection can help scientists to measure the health of this keystone species. We hypothesize that shelled organisms should be more abundant than algae near sea stars because sea stars are more likely to be found in dense areas of food sources, such as mussel beds.

Methods

We measured algae and shelled percentage coverage among 51 *P. ochraceus* and a combined 31 *D. imbricata* and *L. hexactis*. We collected data three separate times at Patrick's Point State Park (41.1298° N, 124.1651° W) and Baker Beach (41.0477° N, 124.1236° W) at low tide during the week of November 10th-17th, 2019 (Fig 1). The low tide levels ranged from -0.3 ft to 2.6 ft. Percent coverage was determined using a 0.50 x 0.50 meter quadrat with 100 squares, each square representing 1% coverage. We centered the quadrat on the sea star and counted squares occupied by algae or shelled organisms. Percent coverage of algae and shelled organisms between *P. ochraceus* and other stars was analysed using the statistical program R. Three tests were run: the Shapiro-Wilk normality test to determine data normalcy, an F-test to compare variances, and a Wilcoxon rank sum test with continuity correction for determining significance of differences seen in the data.

Results

The average quadrat makeup differed between the groups of sea stars we compared. Results from F- and the Shapiro-Wilk test indicated that most data sets did not meet statistical standards for equal variance and normality (Table 1, Table 2); this led us to run a Mann-Whitney test to determine significance (Table 3).

Algae was the most abundant group throughout the duration of the experiment, no matter the asteroid we measured (Fig 2). Even with this, the Mann-Whitney test showed significant differences between algae percent coverage of *P. ochraceus* compared to other sea stars in the intertidal zone; there is significantly more algae surrounding *D. imbricata* and *L. hexactis* than *P. ochraceus* (p-value 2.47e-05, Table 4). When comparing shelled organisms, we see a different pattern arise. The mean percent coverage of shelled organisms is significantly different, but with a majority of shelled organisms surrounding *P. ochraceus* (p-value 0.0004525).

By looking at the data, it is apparent that there was a difference in shelled and algae coverage within each sea star group. There is significantly more algae in both groups compared to shelled organisms (p-value < 2.2e-16, Table 4).

Discussion

Our data and measurements provided our research group with indications that did not support our hypothesis, namely, that if sea stars predominantly feed on shelled organisms, then the percent coverage of shelled organisms would surpass that of algae within each quadrat sample. We recognized that the opposite was true; the percent coverage of algae actually exceeded that of all sea stars and shelled organisms in almost every quadrat that was sampled. Although the experiment we conducted did not cater to our initial hypothesis, the frequency of our measurements provided us with a better understanding regarding the differences, similarities, and patterns between percent coverage of shelled organisms and algae in relation to the different sea stars.

Average percent coverage of algae was greater among *L. hexactis* and *D. imbricata* than that of *P. ochraceus*. However, the average percent coverage of shelled organisms was greater among *P. ochraceus*. This is most likely due to the protection that the algae may provide for smaller sea stars, such as *L. hexactis*; *Leptasterias* spp., which are noted to

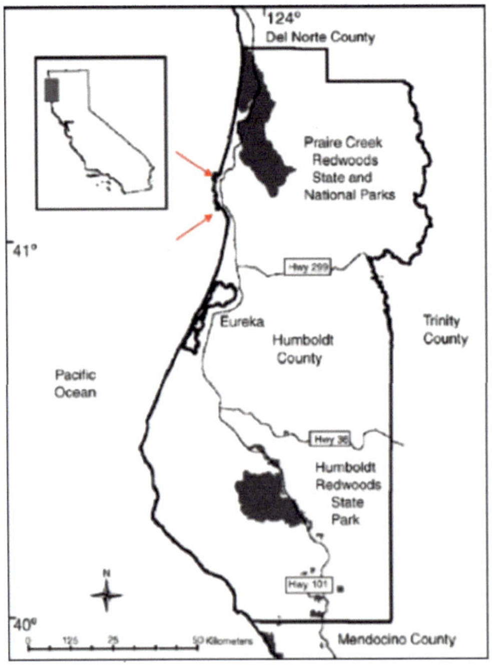

Fig 1. Map of Humboldt County showing the locations of data collection (Baker Beach and Patrick's Point State Park). Adapted from Ashton et al. 2006

reside in mussel beds or in nearby pools, usually in contact with sea grass or algae (Jaffe et al. 2019) We observed that *P. ochraceous* were found more often attached to bare rock, while other sea stars were found more submerged in water. Our data also shows that a variety of other organisms such as anemones, urchins, and nudibranchs were recorded within *L. hexactis* and *D. imbricata* quadrat samples, while *P. ochraceus* was often seen only in the presence of algae or some shelled organisms. The niche of each star ultimately influenced the frequency and diversity of other recorded organisms.

Identifying diet according to each individual species would strengthen this study and may provide more conclusive results in the future, as well as recording data in control quadrats that did not contain sea stars. Understanding the prey selection of sea stars and their location within the intertidal zone drives the understanding of the ecosystem as a whole; a change in sea star behavior would cause catastrophic change in the structure of the intertidal zone. This was observed during the seastar wasting disease outbreak in 2014, when the population of sea stars on the West Coast, specifically *P. ochraceus*, was impacted by the disease. This caused a significant reduction in predation rates of mussel beds (Menge et al. 2016), which results in decreased biodiversity and community structure (Paine 1976). Studying the habitat selection of different sea stars provides future scientists with a basic understanding that is necessary to protect these keystone species in our intertidal zone. With

Table 2. Summary of Shapiro-Wilk test for normality in each data set (W = W-value).

Data sets	W	P-value
Algae (Pisaster)	0.86301	3.549e-05
Algae (Other)	0.93393	0.02672
Shelled (Pisaster)	0.86395	0.0005831
Shelled (Other)	0.81916	0.0007892

Table 3. Summary of Wilcoxon rank sum test comparing differences within the two groups of sea stars and between the two groups of sea star (W= rank totals)

Comparison Groups	W	P-value
Algae: Pisaster v Other	449.5	2.47e-05
Shelled: Pisaster v Other	605	0.0004525
Pisaster: Algae v Shelled	1662	1.255e-13
Other: Algae v Shelled	874	7.621e-11

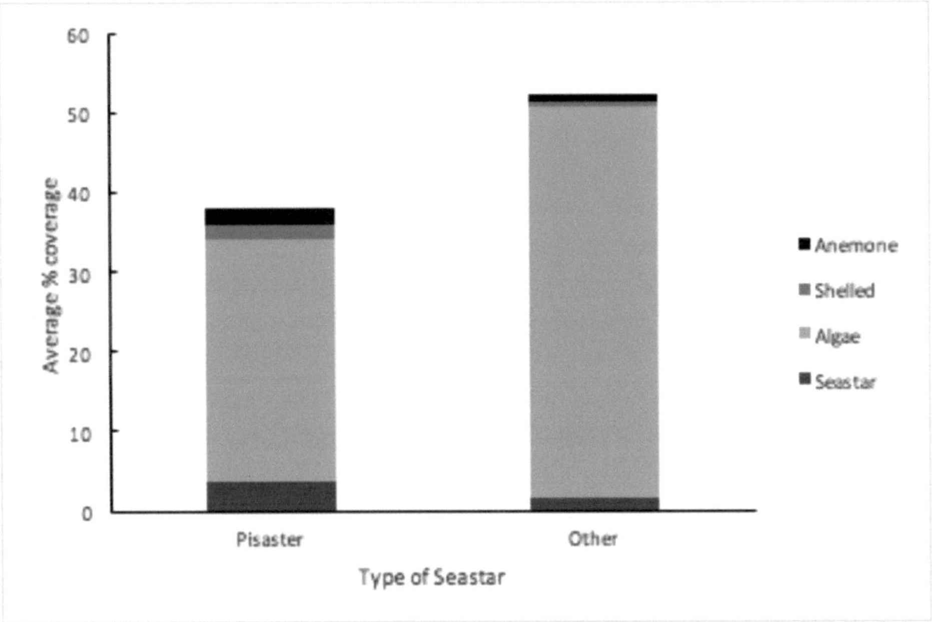

Fig 2. Comparison of mean percent cover of sea star, algae, shelled organisms, and anemones between Ochre stars (*P. ochraceus*) and the Other group (*L. hexactis* and *D. imbricata*)

Table 4. Summary of the averages of algae coverage and shelled organisms organized by type of sea star it was surrounding; group "Other" is a combination of six rayed stars (*L. hexactis)* and leather stars (*D. imbricata*).

Group	*Pisaster ochraceus* Mean	*Pisaster ochraceus* STD	Other Mean	Other STD
Mean % Algae coverage	31	24	49	21
Mean % Shelled organisms	1.9	1.4	0.74	0.62
Mean % Anemone	2.1	1.7	0.80	0.40
Mean % Nudibranch	0	N/A	0.31	0.19
Mean % Chiton	0	N/A	8.0	2.8
Mean % Sea Urchin	0	N/A	6	0

the impending threat of climate change, species higher in the intertidal zone, such as the ones we observed, are at a greater risk of desiccation. The temperatures the sea stars experience determine where in the intertidal zone they are located, which in turn could affect their foraging abilities (Helmuth and Wethey 2009). This stress causes shelled organisms to grow slower, and moving through the food chain would result in less nutrients for the sea stars. The risks of climate change are only growing; knowing how sea stars rely on their prey can help us track the survival of this keystone species.

References

Ashton, Don & Marks, Sharyn & Welsh, Hartwell. (2006). Evidence of continued effects from timber harvesting on lotic amphibians in redwood forests of northwestern California. *Forest Ecology and Management, 221,* 183-193. 10.1016/j.foreco.2005.09.015.

Foster, M. S., et al. (1988). Causes of spatial and temporal patterns in rocky intertidal communities of Central and Northern California. *California Academy of Sciences, 9,* 3-37.

Helmuth, B., & Wethey, D. S. (2009). Climate change in the rocky intertidal zone: Predicting and measuring the body temperature of a keystone predator. *Marine Ecology Progress Series, 374,* 43–56.

Jaffe, Noah et al. (2019). Sea star wasting disease demography and etiology in the brooding sea star *Leptasterias* spp. *PloS one,* 14(11). doi:10.1371/journal.pone.0225248

Menge, Bruce et al. (2016). Sea star wasting disease in the keystone predator *Pisaster ochraceus* in Oregon: insights into differential population impacts, recovery, predation rate, and temperature effects from long-term research. *PLos ONE.* 11(5). doi.org/10.1371/journal.pone.0153994

Manzur, Tatiana et al. (2009). Ontogenetic changes in habitat use and diet of the sea-star *Heliaster helianthus* on the coast of central Chile. *Journal of the Marine Biological Association of the United Kingdom, 90,* 537-546.

Monteiro, A. M. G. & Pardo, E. V. (1994). Food diet of *Astropecten marginatus* and *Luidia senegalensis* (Echinodermata-Asteroidea). *Revista Brasileira de Biologia, 54,* 49-54.

Paine, R. T. (1976). Size-limited predation: An observational and experimental approach with the *Mytilus-Pisaster* interaction. *Ecological Society of America, 57,* 858-873.

Pechenik, J. A. (2015). The Echinoderms. *Biology of the Invertebrates, 2,* 497-527.

Penchaszadeh, P .E. & Lera, M. E. (1983). Diet of 3 tropical *Luidia*-spp Echinodermata Asteroidea in Golfo Triste Venezuela. *Caribbean Journal of Science, 19,* 1-6.

Preliminary Quantitative and Qualitative Investigations on the Heterogeneous, Irreversible Uptake of Acetone by Titanium(IV) Oxide

John S. André (University of Michigan), Natalie Eskandary (Humbolt State University), Terry Franklin (Humboldt State University), Emilia J. McCann (Humboldt State University), Oscar Pascual-Diaz (Humboldt State University), Gregory M. Pitch (University of California, Santa Cruz), Jesus D. Rincon (Humboldt State University), Megan N. Wright (Humboldt State University), Christopher W. Harmon (Humboldt State University)

Keywords: atmospheric chemistry, heterogeneous, uptake, adsorption, desorption, DRIFTS, FTIR

Introduction

Heterogeneous chemical reactions are ubiquitous in the atmosphere and play a strong role in its chemical composition and physical properties.[1-2] Aerosols and/or solid environmental surfaces can alter the chemical composition of the atmosphere by promoting specific reaction pathways that are unique to the surface and their interaction with the gas phase.[3-4] Understanding how heterogeneous chemical reactions alter the physical and chemical properties of the atmosphere is critically important, as these reactions play many roles in the climate system. For example, heterogeneous reactions often increase the hygroscopicity (the capacity of a material to absorb water from the air) of organic aerosols and improve cloud nucleation.[5] The nucleation of clouds by atmospheric aerosols and their effect on light scattering (i.e., the indirect effect) represents one of the largest uncertainties in global climate models.[6-7] Additionally, the direct effect aerosols have on scattering Earth-bound solar radiation also represents a large uncertainty in climate models.[6-7] These reactions can also lead to organic and inorganic aerosols that have deleterious effects on human health.[8] Thus, continued research of heterogeneous reactions on atmospheric aerosols and environmental surfaces is critical as, despite the numerous studies already conducted, many uncertainties in large scale climate models still remain.[7]

The focus of this investigation is the reactive and irreversible uptake of volatile organic compounds (VOC) on environmental metal oxides

(EMO), the principle component of mineral dust. Many studies (see references 3, 4, 10 and references therein) to date have examined the reactivity of EMO and researchers have found them to be important in atmospheric chemistry. For example, long before the heterogeneous chemistry of anthropogenic origins plagued air quality, ice core samples demonstrated that aerosols from mineral dust circled the globe for hundreds of millennia.[9] Mineral dust is largely composed of inorganic compounds, and metal oxides are often the dominant species.[4] Unlike in the chemistry of organic aerosols, metal oxides such as titanium(IV) oxide, TiO_2, are well known for their photocatalytic (the acceleration of a reaction driven by light) and reactive properties. These properties are exploited in environmental remediation,[10] and it has been suggested that these properties could lead to unaccounted-for heterogeneous reactions in atmospheric climate models.[10-11] Indeed, TiO_2 has received much interest in the literature due to its unique catalytic and photocatalytic properties in heterogeneous atmospheric chemistry.[10] Still, despite the large number of studies, many unexplored research directions remain.[10] Conducting quantitative experiments under appropriate conditions of state and composition is the biggest challenge of these; this investigation describes pursuits toward that end.

The central hypothesis of this investigation is that small gas phase VOC can react on mineral dust aerosol surfaces and create larger compounds that are more likely to stick to the surface. These larger compounds would then be transported long distances to regions of the world where they may not be accounted for. The presence of such an organic surface layer will then alter chemical and physical properties of mineral dust aerosols. Indeed, field studies have shown atmospheric mineral dusts are (in addition to complex mixtures of minerals, clays, nitrates, sulfates, etc.[4]) often complex mixtures of various water soluble organic acids,[12] which can form organic salts[13] after being *aged*. We hypothesize the creation of such a layer is condition specific and seek to quantify the formation of such a layer under appropriate conditions of state and composition. Most quantitative information on VOC reactive uptake is conducted under highly idealized conditions of temperature, pressure, and composition. Therefore, we seek to investigate the formation of organic layers on the surface of tailored mixtures of EMO and, eventually, actual dust samples. This investigation will assess the efficacy of mineral dust aerosol to take-up VOC and form an organic layer under actual atmospheric aging conditions of state and composition. Preliminary qualitative and quantitative work is presented here using a baseline condition with a representative VOC (acetone) and EMO surface compound (TiO_2).

Mineral dust is eroded soil particles that have been become aloft in the atmosphere. They are the second largest contributor to the global aerosol burden, after sea spray aerosols.[14] The chemical composition of mineral dust is very similar, if not identical, to that of the crustal rock from which the dust originated.[15] Many locations around the world report mineral dust to be fairly uniform in composition (% by weight) with respect to SiO_2 (~60%) and Al_2O_3 (~10-15%). The lower abundance oxides, such as CaO (~6%), Fe_2O_3 (~5%), MgO (~4%), and TiO_2 (~1%), are variable depending on location.[16,15] Most notably and pertinent to this study is TiO_2, which despite having the lowest abundance is generally the most reactive due to its photocatalytic properties[10] and has the largest uptake coefficient (the probability that a molecular gas-surface collision leads to uptake of the gas by the solid) in comparison to those EMOs listed above.[17]

Carbonyl compounds are of interest in this study as they are among the most prevalent and reactive VOC in the atmosphere, arising mostly from photo-oxidation of their parent hydrocarbon.[10] For example, a field study identified the three most common (of 23) carbonyl compounds found in the Los Angeles troposphere to be formaldehyde, acetaldehyde, and acetone, with total carbonyl four-hour mixing ratios of 22–29 parts per billion by volume (ppbv).[18] Shown in Figure 1 is an example of a reaction that can occur through the EMO or acid initiated catalysis of VOC: an

Aldol condensation reaction occurring with carbonyl compounds to form larger molecular weight compounds.[17, 19-21] If the starting carbonyl is acetone, first mesityl oxide (an enone) can form and further reaction can lead to isophorone.[17, 19-21] Additionally, isophorone is known to undergo photodimerization under actinic wavelengths (~300 nm)[22] in aqueous solutions (not shown in Figure 1), leading to even larger surface compounds. Thus, the creation of an organic layer on EMO surfaces under atmospheric conditions is very possible.

An interesting observation in this reaction scheme is the creation of larger compounds via the formation of new carbon-carbon double bonds. Olefins are quite reactive in the atmosphere and are subject to oxidation via ozone or the hydroxyl radical. This suggests the reaction/irreversible uptake of carbonyl compounds on EMO could alter the chemical properties of mineral dust because new carbon-carbon double bonds on the surface will be a sink for ozone. These reaction schemes are also fundamentally interesting, as the O:C (oxygen to carbon) ratio is *decreasing* in contrast to the fundamentally accepted view that O:C ratio should increase with chemical aging in the atmosphere.[5] This also suggests such an organic layer could alter the physical properties of mineral dust because water uptake could be hindered with a decrease in O:C.[5]

The primary results presented here deal with sparingly small surface coverage layers (Θ, defined as the number of monolayers) of acetone being adsorbed (i.e., stuck) or desorbed (i.e., unstuck) from a TiO_2 surface. These investigations will establish the initial stages of testing our central research hypothesis described above, where quantitative information on the adsorption and desorption rates are presented. Additionally, qualitative investigations on the reaction of acetone over TiO_2 to form mesityl oxide (Figure 1) will also be addressed, as well as future directions of research.

Methods

The Diffuse Reflectance Infrared Fourier Transform Spectroscopy (DRIFTS) technique will be the principle analytical instrument used in this research, which has been used extensively in atmospheric chemistry research.[2] In brief, a powdered solid is loaded into a reaction chamber (inside the sampling compartment of a Fourier Transform Infrared Spectrometer, FTIR), where the gas phase environment over the surface of the sample can be manipulated. The sample chamber is outfitted with infrared (IR) transparent windows (Zinc Selenide, ZnSe) and the analysis beam is scattered diffusely off the sample surface, focused by a set of mirrors, and then analyzed by a detector. This allows a reaction to be conducted *in-situ* and to monitor changes to the surface in real time. The IR beam will only analyze the surface of the sample, typically about 1–10 μm deep on non-absorbing samples.[23] Challenges associated with the DRIFTS technique are related to particle size and uniformity as well as to a very low signal that is scattered diffusely off the surface. This makes using the technique quantitatively difficult. However, as long as particle sizes are in the micron range and don't change size significantly throughout a reaction, the technique can be used quantitatively.[24] Thus, the technique is ideal because we wish to use particles in the micron size range, which are more relevant to mineral dust in the atmosphere. Additionally, highly reflective (weakly absorbing between 4,000-400 cm^{-1}) powders are ideal for DRIFTS,[24] which EMO particles most certainly are.

Shown in Figure 2 is the experiment station for studying the reactive/irreversible uptake of VOC with EMO via the DRIFTS technique. This entire apparatus (as well as the majority of data presented here) was built and tested by undergraduate research students under the supervision and guidance of the principle investigator (PI).[25] Pictured in the top left corner of Figure 2 is the experiment station located on the 3rd floor of the Science A building on the Humboldt State University campus, where the Chemistry Department is located. The station is organized in three main areas: the FTIR, the manifold, and the computer for data acquisition.

The manifold is built in an instrument rack and is capable of being relocated, if necessary. In the manifold, gas flow is controlled via an MKS

Instruments Type 247D 4 channel readout and the corresponding MKS mass flow controllers (MFC – M100 B) of differing ranges, shown in Figure 2. All gases are UHP (ultra-high purity, >99.999% Airgas®) and are delivered via ¼" PTFE tubing with various Swagelok® connections. Channel 1 (MFC 1) is predominantly used for O_2, channel 3 (MFC 3) is predominantly used for N_2, and channel 2 (MFC 2) can use He or N_2, and all are controlled by switch valves depending on the experiment. Ozone (O_3) can be generated, if desired, in the manifold as depicted in Figure 2. A stream of gas-phase VOC is created by diluting the flow of an organic liquid from a syringe pump (KD Scientific, KDS-100) with the flow of UHP N_2 from MFC 3 (Figure 2). A 25 µL Hamilton syringe (gas-tight) is used with a septum inserted into a Swagelok® union tee and can generate reliable flow rates of 0.1–10 µL/hr. When a VOC is mixed with 100 SCCM (standard cubic centimeters per minute) of N_2 from MFC 3 at a syringe pump flow rate of 0.1 µL/hr, a stable mixing ratio of a nominal organic (molar mass ~75 g/mol, density ~0.8 g/mL) of 4 parts per million by volume (ppmv) can be generated after traversing the 25-foot mixing coil (Figure 2).

The TiO_2 samples were prepared for analysis by DRIFTS using a similar version of the grinding method described by Finlayson-Pitts et al.,[26-27] which is known to create fairly uniform particle sizes between 1–5 µm. Preliminary imaging studies have been conducted as The College of Natural Resources and Science (CNRS) instrument core contains a Scanning Electron Microscope (SEM, Quanta 250) available to properly trained undergraduate students. Figure 3 shows one example, which demonstrates that, after grinding, the particles are polydisperse in size with some being greater than ~1 µm and most being in the sub-micrometer regime. After grinding, the TiO_2 is stored in a crucible in an oven at 115 °C and is only exposed to atmosphere for very brief periods of time during loading into the spectrometer. In the quantitative studies presented here, five nearly identical samples were created, using the methods described above, to investigate the reproducibility of our methodologies. These samples where loaded into shallow (~3 mm deep) cylindrical sample cups with a surface area exposed to the gas phase, ideal for analysis via DRIFTS, where the physical properties of these samples are shown in Table 1 below.

TABLE 1: Physical Properties of TiO_2 Samples

Sample	Mass (mg)	Exposed Geometric Surface Area (mm^2)
1	103.6	46.6
2	133.7	47.8
3	128.1	51.5
4	99.7	45.4
5	129.1	46.6

Pictured in the top right corner of Figure 2 is the DRIFTS cell (Spectra Tech Inc., 6106640), which consists of a flow-through manifold, vacuum line, heater block, thermocouple, and manifold for liquid coolant. All preliminary experiments have been conducted without heating or cooling the sample, at ~28°C. The sample holder is outfitted with ZnSe optics to minimize damage due to exposure of ozone or water vapor, which can degrade salt optics such as KBr. Thus, the DRIFTS cell can maintain a desired environment over the surface of the sample. In a typical experiment, a mixture of gases from the manifold is delivered to the DRIFTS cell at 100. standard cubic centimeters per minute (SCCM) to avoid pressurizing the cell (DRIFT cell volume = 5.3 cm^3, 1/16" Swagelok® connections), which is confirmed by measuring the flow rate before and after the cell. A flow of UHP purge gas air (Parker-Hannifin Spectra 30) is also delivered to the FTIR (Thermo Nicolet Inc., Avatar 370) continuously whether experiments are being conducted or not.

Uptake studies are initiated by recording a background single beam spectrum of TiO_2 after purging the sample with 100. SCCM of UHP N_2 for ~1 hour. Subsequent spectra are then recorded by ratioing new single beam spectra to the initial single beam spectrum using the Kubelka–Munk (KM) algorithm (preferred qualitative methodology)[28] or the standard Absorbance (A) algorithm (preferred quantitative methodology)[2] after a change is imposed on the surface (i.e., uptake of VOC). Using this methodology, the spectra reveal dynamic processes occurring at the gas/solid interface.[26-28] Negative features in the spectra indicate that a functional group has left the surface or has been modified, whereas positive features in the spectra show that a new functional group has been created on the surface or an existing functional group has been modified.[26-28] A representative series of spectra is shown in Figure 2, where different colors indicate evolving oxidation times in a different experiment not discussed in this article. In nearly all experiments, anywhere from 100–250 scans are acquired and averaged into one spectrum with a resolution of 4 cm^{-1} using an intermediate interferometer speed (0.47 cm/s) and the largest gain, which maximizes instrument sensitivity but may degrade the signal to noise ratio.

Results

Shown in Figure 4 is KM DRIFTS spectra of the adsorption of 5 ppmv of acetone (purified by distillation) in the gas phase (N_2 only) on TiO_2. Spectra were recorded as described above, where acetone adsorption spectra were ratioed to TiO_2 that was purged with dry UHP N_2 for one hour. Table 2 also summarizes the results of acetone adsorbed to TiO_2, which are consistent with previous studies.[19-21, 29] Not shown in Table 2 is the negative peak at 3670 cm^{-1}, shown in Figure 4(c), which corresponds to modification of surface bound H_2O and OH groups on TiO_2 as a result of acetone adsorption. This peak is consistent with other observations[19-21, 29] and is a good indication that the spectra, shown in Figure 5, arise from irreversibly, adsorbed acetone.

Shown in Figure 5 is a representative quantitative experiment (Sample #5), where spectra were recorded in Absorbance mode. The experimental data shown in Figure 5 was acquired as described above, where a series of single beam spectra are ratioed at various time intervals as acetone is flowing over the surface of TiO_2. The manifold has a bypass 3-way valve (not pictured in Figure 2), where a stable mixing ratio of acetone can be established and exhausted to a fume hood prior to exposure to the sample surface in the DRIFTS cell. This allows us to initially establish a clean and dry single beam background spectrum and switch the bypass valve from *exhaust mode* to *exposure mode*, which establishes the beginning of a kinetic experiment (i.e., *time* = 0). Adsorption studies are initiated by switching from exhaust mode to exposure mode and are conducted for ~1 hour using approximately 50 ppmv of acetone, after which time desorption studies are initiated by switching the valve back to exhaust mode. Thus, each time an adsorption experiment was conducted a desorption experiment followed immediately, where the sample is purged with UHP N_2 only and ratioed to the same initial single beam background prior to any acetone exposure.

Each of the samples shown in Table 1 were tested twice, where an adsorption study was initiated, as described above, followed immediately by

TABLE 2: Peak Assignments of Acetone Adsorbed on TiO_2

Vibration Mode	Frequency (cm^{-1}) This Work	Frequency (cm^{-1}) Literature	References
ν(C–H)	2972	2971–2973	19-21, 29
ν(C=O)	1703, 1693	1688–1702	19-21, 29
ν(C–C)	1240	1238–1243	19-21, 29
$\delta_{as}(CH_3)$	1423	1420–1422	19-21, 29
$\delta_s(CH_3)$	1365	1365–1368	19-21, 29

a desorption study. Interestingly, all samples tested (and their respective trial replicates) showed very similar behavior to that seen in Figure 5 with the exception of sample #2, which did not demonstrate any level of acetone adsorption. While the behavior of each sample (less #2) was very similar, with respect to adsorption/desorption of acetone, the absolute absorbance data was not identical. For example, Figure 6(a) shows the first trial replicate peak absorbance for samples #1, 3, 4, & 5 in adsorption mode (note, sample #2 is not shown as no adsorption or desorption was detected). As shown in Figure 6(a), the *absolute* peak absorbance from sample to sample is not the same; however, when the data is *normalized* the adsorption and desorption data from sample to sample is very consistent, shown in Figure 6(b). This was done by normalizing the area under the curve between 1740–1660 cm^{-1} (i.e., the carbonyl region) in spectra, recorded just prior to switching to exhaust/desorption mode. Error bars arise in this analysis from also averaging in the normalized peak absorbance. In this methodology, all experiments will achieve adsorption equilibrium at a normalized value of 1.0 and the desorption equilibrium value is unknown *a priori*. Predictive models for adsorption and desorption are also shown as solid lines and shaded areas in Figure 6(a,b), which will be discussed in more detail below.

When a large surface density of acetone molecules are present ($\Theta > 0.3$)[19-21, 29], they are in close enough proximity to each other to undergo an Aldol condensation, leading to the formation of mesityl oxide and possibly isophorone as shown in Figure 1. This result has been reasonably corroborated and is shown in Figure 7. In this experiment a large (qualitative, unknown *a priori*) amount of acetone was passed over the surface of TiO_2 in the DRIFTS cell for ~60 minutes, where similar behavior as that seen in Figure 5 was observed. However, when the experiment was switched into desorption mode (by purging the surface with UHP N_2) the spectra revealed a chemical change as seen in Figure 7(a,b).

The classic text-book[30] acid-catalyzed aldol condensation reaction mechanism is shown in Figure 7(c) to assess if the spectra shown in Figure 7(a,b) are evidence of mesityl oxide formation. In step 1 of the mechanism, acid catalyzed enolization occurs via protonation of the carbonyl group, leading to the production of water and an electrophilic protonated ketone (i.e., the conjugate acid of acetone). Some of those conjugate acids can undergo further proton transfer reactions with available water molecules to create a stable enol, which will act as a nucleophile (i.e., Lewis base) and attack the carbonyl carbon of the electrophilic protonated ketone shown by step 2. The first two steps in the mechanism are critical, as they are what create new carbon-carbon bonds, owing to the creation of larger molecular weight compounds. The evolution of function groups in DRIFTS spectra from 0–17 minutes suggests this mechanism (or something similar) is occurring on the surface of the TiO_2 particles. As can be seen in Figure 7(a), from 0–7 minutes the surface hydroxyl peak (3670 cm^{-1}) reduced in intensity while a new band most likely associated with water (ν_s/ν_{as}, 3412 cm^{-1}) appeared. This is evidenced with the protonation of the ketone group and production of water in step 1. Additionally, from 0–7 minutes Figure 7(b) shows a drastic reduction in the intensity of the carbonyl peak (ν(C=O), 1703 cm^{-1}) with the appearance of a shoulder at 1666 cm^{-1}. As the spectra evolve from 7–17 minutes the carbonyl peak completely disappears, consistent with the nucleophilic attack of the enol on the conjugate acid leading to the creation of an α-hydroxy carbocation shown by step 3. Additionally, ν(C–H) and $\delta(CH_3)$ features (3,000–2800 and 1447 cm^{-1}, respectively) are seen during this time interval, which is also consistent with steps 2 and 3. As the spectra evolve from 17–57 minutes it becomes difficult to discern the mechanism and associated functional groups. For example, previous researchers have assigned adsorbed mesityl oxide ν(C=C) and ν(C=O) to be 1602 and 1666 cm^{-1}, respectively[29], which we do see in our spectra; however, the intensity of both of these peaks should be very similar. Figure 6(d) shows a NIST reference spectra of mesityl oxide vapor,[31] where the intensities of both ν(C=C) and ν(C=O) are similar (albeit shifted ~30 cm^{-1} higher due to it being a gas). Indeed, previous researchers also see similar intensities in both ν(C=C) and ν(C=O) for adsorbed mesityl oxide.[29] Thus, it is unclear if these spectra are in fact confirmation of mesityl oxide formation or if we are just limited by an older, less sensitive instrument. Similarly, the NIST reference spectra of isophorone vapor[31] is also shown, which is a potential product that could form through further acid catalyzed (or surface catalyzed) Aldol condensation reaction mechanisms.[17] The ν(C=O) peak at 1694 cm^{-1} (likely shifted to around ~1660 cm^{-1} in the adsorbed phase) should be much more prominent than the ν(C=C), which does not appear to be the case in the DRIFTS spectra shown in Figure 7(a,b) either. These observations raise the question if they are from a limitation of our instrument or actual differences in the observed chemistry. Further investigations are most certainly warranted.

Discussion

The combination of qualitative and quantitative data, shown above, represents a baseline level of progress towards investigating our central hypothesis: that small gas phase VOC can stick and react on mineral dust aerosol surfaces, which will create larger compounds that can modify the physical and chemical properties of such mineral dust. The efficacy at which these VOC adsorb (i.e., stick) and desorb, we predict, is highly dependent on the surface conditions present in a sample, such as its surface area, crystal structure, what is has been exposed to prior to VOC exposure, etc. Thus, a number of quantitative parameters can be used to compare each of these condition specific adsorption/desorption profiles; however, 1st order rate constants for adsorption/desorption will be the principle quantitative parameter that is presented for now.

A simple 1st order kinetic model approaching equilibrium was fit to the data, shown by equation 1, where $[A]_t$, $[A]_e$ and $[A]_0$ can represent the normalized or un-normalized concentrations at any time (t), equilibrium, and initially ($t=0$), respectively, k is the first order rate constant, and t is time in minutes.

$$[A]_t = [A]_e + ([A]_0 - [A]_e)e^{-kt} \quad (1)$$

This model can be applied to the adsorption or desorption studies, where in the adsorption studies the theoretical normalized value for $[A]_e$ and $[A]_0$ are 1 and 0, respectively, thus making equation 1: $1-e^{-kt}$. In the desorption studies the theoretical values for $[A]_e$ are unknown *a priori* and that of $[A]_0$ should be 1. These predictive models are shown as solid lines in Figure 6(a) for each respective sample (where values are not normalized) as well as for the entire normalized data set in Figure 6(b). In each of these parameterizations the rate constants, k, should be very similar; however, the equilibrium and initial values ($[A]_e$ and $[A]_0$, respectively) will obviously be different whether the data has been normalized or not.

TABLE 3: Summary of fit parameters[a] given by equation 1

Sample	Adsorption k (min^{-1})	Adsorption $[A]_e$	Adsorption $[A]_0$	Desorption k (min^{-1})	Desorption $[A]_e$	Desorption $[A]_0$
1	0.124±0.023	0.958±0.029	0.013±0.090	0.0405±0.0230	0.741±0.036	0.965±0.216
3	0.117±0.018	0.968±0.035	0.006±0.073	–	–	–
4	0.133±0.043	1.042±0.060	0.026±0.152	0.0436±0.0161	0.669±0.042	0.983±0.146
5	0.112±0.018	1.006±0.033	-0.032±0.069	0.0503±0.0223	0.629±0.054	0.998±0.181
All	0.116±0.014	0.957±0.022	0.032±0.056	0.0631±0.0262	0.716±0.029	0.993±0.171

[a] Fit parameters are conducted at the 95% confidence interval

Summarized in Table 3 are results for the 1st trial replicate of the normalized results and their respective quantitative parameters (note, due to technical errors the desorption data for Sample 3 was not able to be analyzed; however, did proceed very similarly as the other data sets).

The results are presented as individual samples as well as for all data (i.e., "All"), which is the exact same data that is seen in Figure 6(b). As can be seen in Table 3, the theoretical values for $[A]_e$ and $[A]_{0,}$ with respect to adsorption, are as exactly as they should be, barring some amount of data fitting error associated with the very low signal present in spectra. Additionally, all rate constants for both adsorption and desorption studies are self-consistent with each other and within the 95% confidence interval.

Most interesting for this study is the difference between adsorption and desorption rate constants, where, consistently, the adsorption rate constant is nearly double that of the desorption rate constant. Applying fundamental concepts of equilibrium to this observation, a simple equilibrium constant (K_{eq}) and corresponding change in Gibbs Free Energy (ΔG) can be calculated, which is shown by equation 2.[32] Here k_{ad} and k_{des} represent the adsorption (i.e., *forward*) and desorption (i.e. *reverse*) rate constants, respectively, R is the universal gas constants in units of J mol^{-1} K^{-1}, and T is the temperature in units of K.

$$\frac{k_{ad}}{k_{des}} = K_{eq} = e^{-\Delta G/RT} \qquad (2)$$

Using this equation K_{eq} is calculated to be (for ALL data) 1.84 ± 0.80 and ΔG (at 287 K, the average planet-wide temperature) is −1.51 ± 0.66 kJ mol^{-1}. Despite the fact that there is fairly large error in these parameters (~40%), the sign and magnitude are significant and confirm our central hypothesis that small VOC can stick to the surface of EMO. For example, a negative Gibbs Free energy (correspondingly, an equilibrium constant greater than 1) suggests acetone will spontaneously stick to an EMO surface at most atmospheric temperatures.[32] This is confirmed by the fact that the equilibrium parameter value (i.e., $[A]_e$) in the desorption studies is 0.716 ± 0.029, which means, on average, after acetone has spontaneously stuck itself to a surface (i.e., adsorption) roughly 68.7–74.5% will remain on the surface at equilibrium. Indeed, to test this observation all samples were left in the DRIFTS sample chamber overnight, after the ~1 hour of desorption studies, and 24-hours later the acetone signal was very similar, if not identical, to the last spectra recorded! This indeed confirms acetone attains an equilibrium surface coverage and will NOT reversibly desorb.

Conclusions and Future Directions

In this study we have shown that acetone, a VOC largely present in the atmosphere, can spontaneously and irreversible stick to a TiO_2 surface, which is a small, yet representative component of mineral dust aerosol. This *sticking* appears to be quite strong and, over atmospheric time scales of hours and days, may remain stuck to the mineral dust surface and be transported long distances to another part of the world, where it is perhaps not accounted for. Additionally, when the surface coverage of acetone is large (and unknown *a priori* in our experiments) it self-reacts via the Aldol Condensation mechanism and forms a larger compound that could change the surface properties of its host mineral dust.

The preliminary work we have presented here is significant to the atmospheric chemistry and air pollution community as we have shown small gas phase molecules, which are expected to remain in the gas phase, can stick irreversibly to solid aerosol (specifically mineral dust) surfaces. Given that the direct and indirect effects aerosols have on scattering Earth-bound solar radiation represents a large uncertainty in climate models[6-7], continued research on aerosols and their interaction with the atmosphere is desired. For example, the reactions shown here can modify the chemical and physical properties of mineral dust, making their role in the climate system less known. Additionally, adsorbed surface organic compounds (i.e., acetone, mesityl oxide, etc.) can also lead to aerosols that have deleterious effects on human health.[8] Thus, continued research in this direction is critical as, despite the numerous studies already conducted, many uncertainties in large scale climate models and the effect atmospheric aerosols have on human health still remain.[7]

In the future, we wish to systematically *age* our EMO samples prior to any VOC adsorption/desorption by exposing those EMO samples to light, relative humidity, an oxidizing environment, etc., and measure how the adsorption/desorption rate constants with respect to acetone (or any other VOC) change as a result. These future experiments aim to mimic actual chemical and physical aging as it may occur in the real environment/atmosphere. Additionally, we also seek to quantify how the formation of mesityl oxide changes as a result of this *atmospheric aging*. We already have some evidence that suggests these types of surface processing or aging can make a big difference in the uptake of VOC. For example, Sample #2 consistently did not display any type of irreversible uptake or surface chemistry despite the fact that it was created identically to the other samples and looks identical as well. A simple hypothesis for this observation is that the surface of Sample #2 was inadvertently exposed or contaminated to some laboratory or atmospheric species, thus rendering its surface chemistry inert. Further investigation is most certainly desired. Last, we seek to gain even more quantitative information by gaining the ability to quantify the absolute number of molecules present on the surface. This will allow us to investigate the kinetic competition between irreversible adsorption/desorption and formation of mesityl oxide, which will be critically important to validate our hypothesis that the presence of large surface compounds can change the physical and chemical properties of the host mineral dust material. Investigating how the physical and chemical properties of mineral dust are altered by the presence of such surface compounds, is also a future direction of research in our group.

Figures

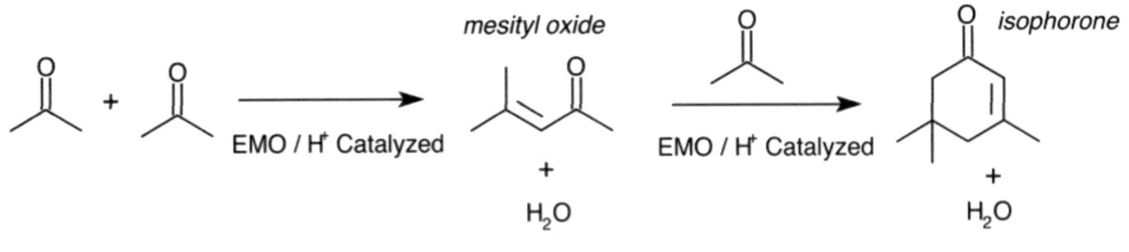

Figure 1: Acetone Aldol condensation occurring through EMO or acid catalyzed mechanisms.

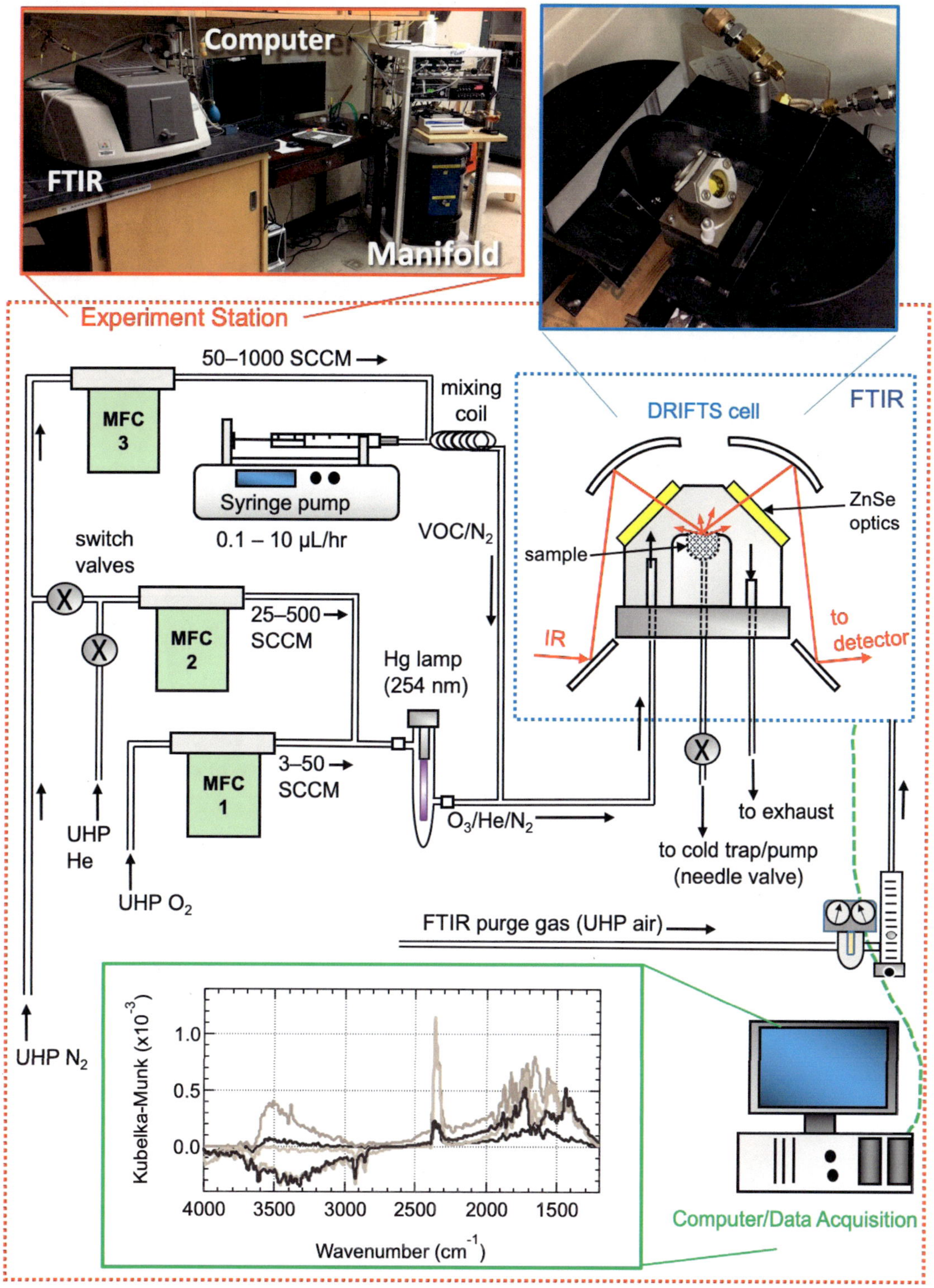

Figure 2: Experimental DRIFTS and manifold station. Key: VOC = volatile organic compound, MFC = mass flow controller, UHP = ultra-high purity, SCCM = standard cubic centimeter per minute, DRIFTS = diffuse reflectance infrared Fourier transform spectroscopy, IR = infrared.

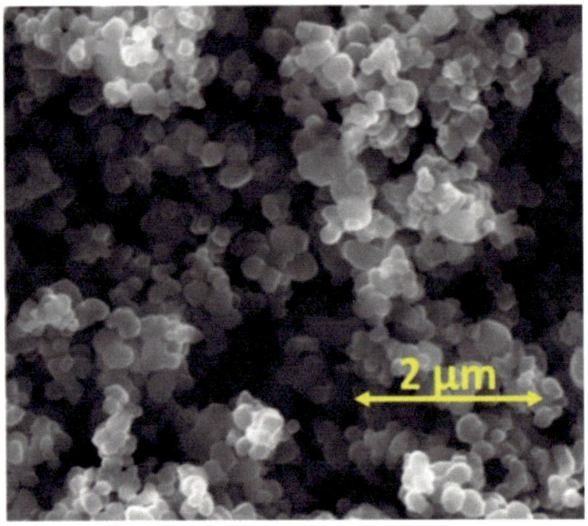

Figure 3: SEM Image of TiO_2 particles.

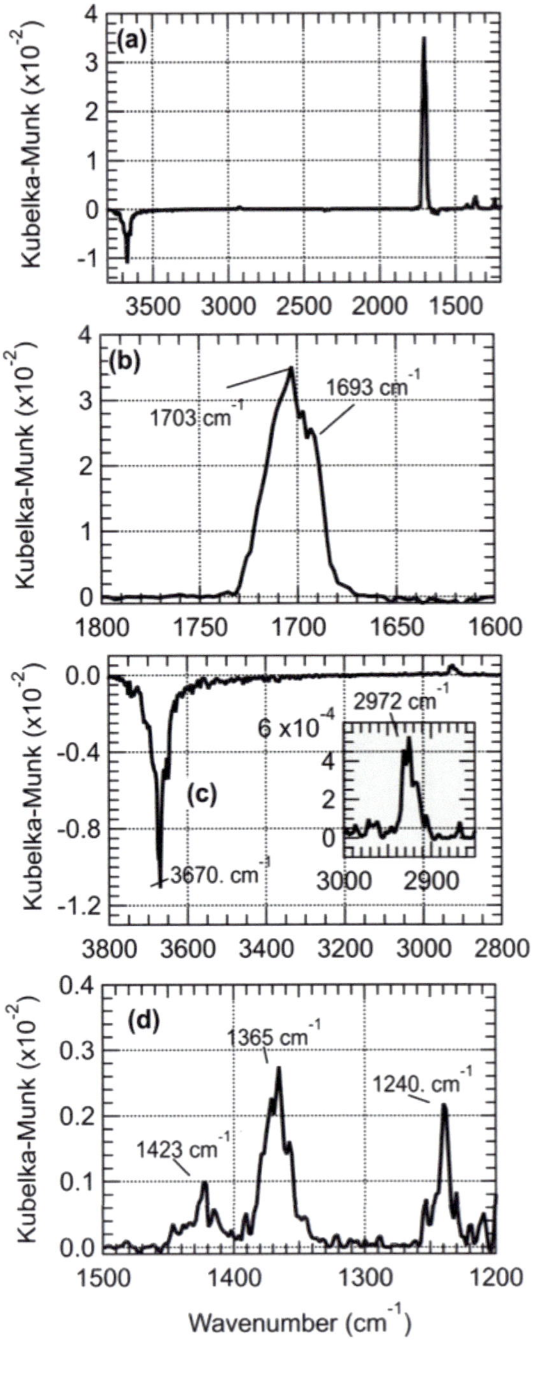

Figure 4(a): DRIFTS spectrum of adsorbed acetone (5 ppmv gas phase) on TiO_2. **(b)** carbonyl region, **(c)** O-H and C-H stretching region, **(d)** C-C and C-H region

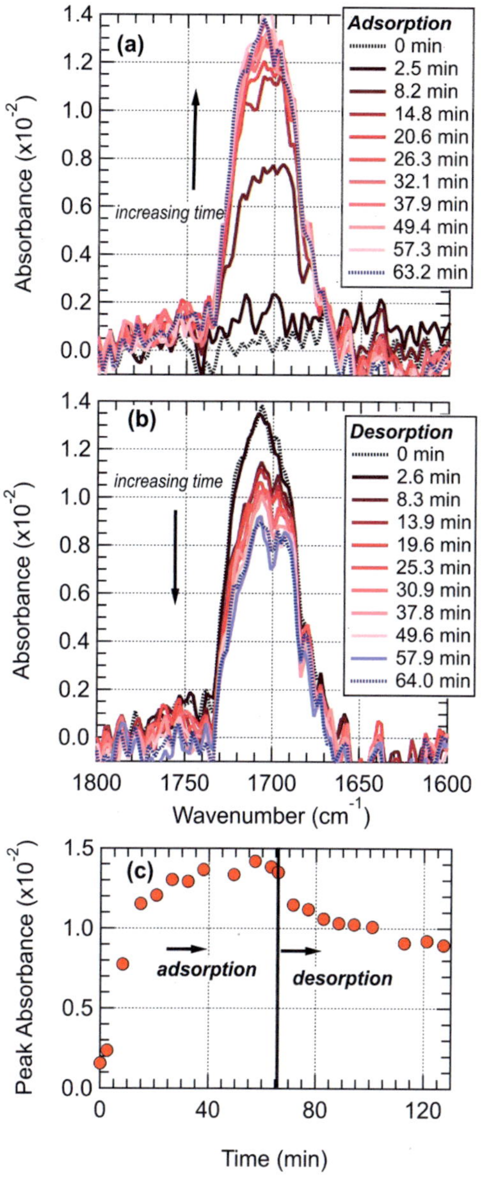

Figure 5(a): DRIFTS spectrum of adsorbed acetone (50 ppmv gas phase) on TiO_2 as a function of time. **(b)** Desorption of acetone (50 ppmv gas phase) on TiO_2 as a function of time. **(c)** Peak absorbance (~1700 cm^{-1}) as a function of total time (note, 63.2 min on adsorption graph = 0 min on desorption graph).

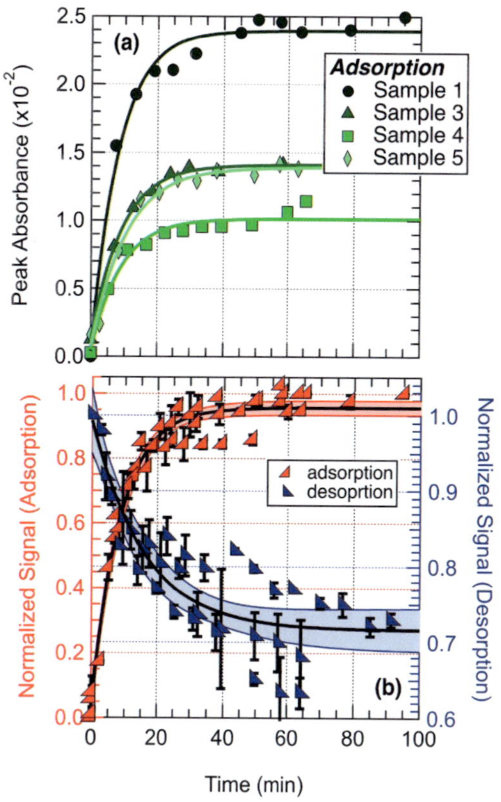

Figure 6: **(a)** Peak absorbance of adsorbed acetone (50 ppmv gas phase) on TiO_2 as a function of time and sample. **(b)** Normalized absorbance signal (adsorption and desorption) as a function of time for Samples 1, 3, 4, & 5. In both **(a)** and **(b)** the data points are measured quantities whereas the solid lines (and shaded areas) are predictive models.

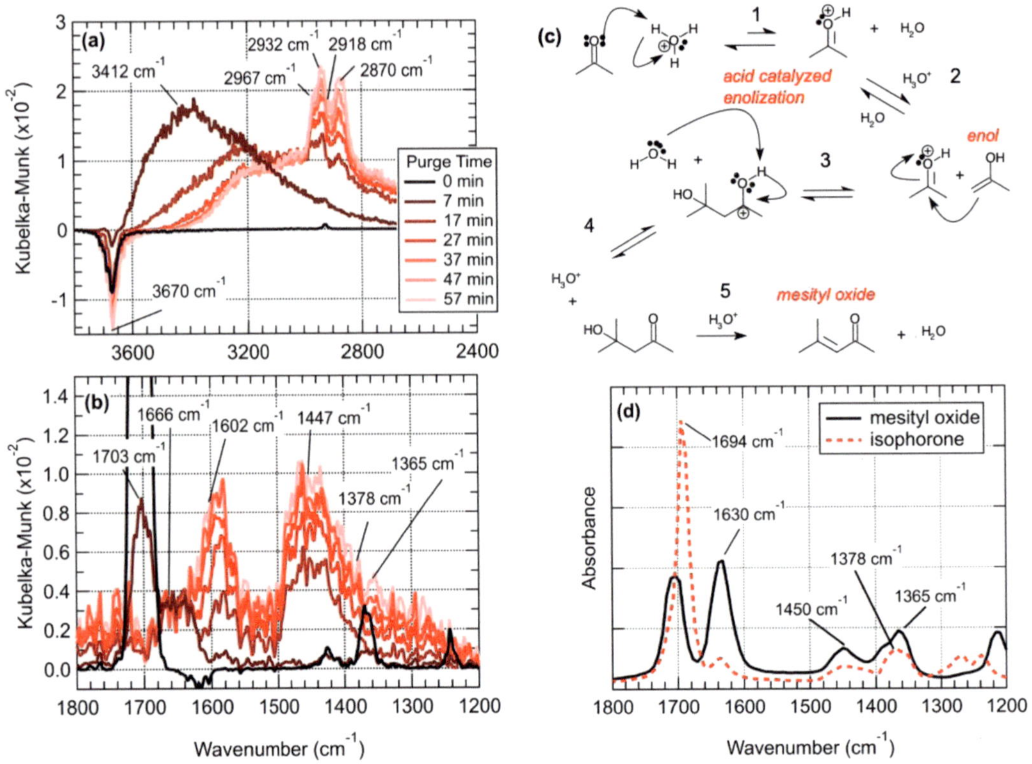

Figure 7(a,b): DRIFTS spectra of acetone reacting on a TiO_2 surface. **(c)** Text book acid-catalyzed Aldol condensation reaction. **(d)** NIST reference spectra of mesityl oxide and isophorone.

References

1. Finlayson-Pitts, B. J.; Pitts, J. N., *Chemistry of the Upper and Lower Atmosphere: Theory, Experiments, and Applications*. Academic Press: San Diego, 2000; p 1040 pp.
2. Finlayson-Pitts, B. J., Reactions at surfaces in the atmosphere: integration of experiments and theory as necessary (but not necessarily sufficient) for predicting the physical chemistry of aerosols. *Physical chemistry chemical physics : PCCP* **2009,** *11* (36), 7760-79.
3. George, C.; Ammann, M.; D'Anna, B.; Donaldson, D. J.; Nizkorodov, S. A., Heterogeneous photochemistry in the atmosphere. *Chem. Rev.* **2015,** *115* (10), 4218-58.
4. Usher, C. R.; Michel, A. E.; Grassian, V. H., Reactions on Mineral Dust. *Chem. Rev.* **2003,** *103* (12), 4883-4939.
5. Harmon, C. W.; Ruehl, C. R.; Cappa, C. D.; Wilson, K. R., A statistical description of the evolution of cloud condensation nuclei activity during the heterogeneous oxidation of squalane and bis(2-ethylhexyl) sebacate aerosol by hydroxyl radicals. *Physical chemistry chemical physics : PCCP* **2013,** *15* (24), 9679-93.
6. Myhre, G.; Shindell, D.; Bréon, F.-M.; Collins, W.; Fuglestvedt, J.; Huang, J.; Koch, D.; Lamarque, J. F.; Lee, D.; Mendoza, B.; Nakajima, T.; Robock, A.; Stephens, G.; Takemura, T.; Zhang, H., Anthropogenic and Natural Radiative Forcing. In *Climate Change 2013:*

The Physical Science Basis. Contribution of Working Group I to the Fifth Assessment Report of the Intergovermental Panel on Climate Change, Stocker, T. F.; Qin, D.; Plattner, G.-K.; Tignor, M.; Allen, S. K.; Boschung, J.; Nauels, A.; Xia, Y.; Bex, V.; Midgley, P. M., Eds. Cambring University Press: Cambrige, UK and New York, NY, USA, 2013.
7. IPCC, 2013: Summary for Policymakers. In *Climate Change 2013: The Physical Science Basis. Contribution of Working Group I to the Fifth Assessment Report of the Intergovermental Panel on Climate Change*, Stocker, T. F.; Qin, D.; Plattner, G.-K.; Tignor, M.; Allen, S. K.; Boschung, J.; Nauels, A.; Xia, T.; Bex, V.; Midgley, P. M., Eds. Cambridge University Press: Cambridge, UK and New York, NY, USA.
8. Samet, J. M.; Zeger, S. L.; Dominici, F.; Curriero, F.; Coursac, I.; Dockery, D. W.; Schwartz, J.; Zanobetti, A. *The National Morbidity, Mortality, and Air Pollution Study. Part II: Morbidity and mortality from air pollution in the United States*; Health Effects Institute: 2000; pp 5-70.
9. Kohfeld, K. E.; Harrison, S. P., DIRTMAP: the geological record of dust. *Earth-Sci. Rev.* **2001,** *54* (1–3), 81–114.
10. Chen, H.; Nanayakkara, C. E.; Grassian, V. H., Titanium dioxide photocatalysis in atmospheric chemistry. *Chem. Rev.* **2012,** *112* (11), 5919-48.
11. Parmon, V. N.; Zakharenko, V. S., Photocatalysis and Photosorption in the Earth's Atmosphere. *Catt* **2001,** *5* (2), 96–115.
12. Lee, S.-H.; Murphy, D. M.; Thomson, D. S.; Middlebrook, A. M., Chemical components of single particles measured with Particle Analysis by Laser Mass Spectrometry (PALMS) during hte Atlanta SuperSite Project: Focus on organic/sulfate, lead, soot, and mineral particles. *JGR* **2002,** *107* (D1).
13. Laskina, O.; Young, M. A.; Kleiber, P. D.; Grassian, V. H., Infrared spectroscopy and micro-Raman spectroscopy of select compoenents of mineral dust mixed with organic coating. *Journal of Geophysical Research Atomspheres* **2013,** *118*, 6593–6606.
14. Huneeus, N.; O., B.; Chevallier, F., Atmospheric Inverstion of SO2 and Primary Aerosol Emissions for the year 2010. *ACP* **2013,** *13*, 6555–6573.
15. Goudie, A. S.; Middleton, N., Saharan Dust Storms: Nature and Consequences. *Earth-Sci. Rev.* **2001,** *56*, 179–204.
16. Wedepohl, K. H., The composition of the continental crust. *Geochim. Cosmochim. Acta* **1995,** *59* (7), 1217–1232.
17. Li, P.; Perreau, K. A.; Covington, E.; Song, C. H.; Carmichael, G. R.; Grassian, V. H., Heterogeneous reactions of volatile organic compounds on oxide particles of the most abundant crustal elements: Surface reactions of acetaldehyde, acetone, and propionaldehyde on SiO2, Al2O3, Fe2O3, TiO2, and CaO. *JGR* **2001,** *106* (D6), 5517-5529.
18. Grosjean, E.; Grosjean, D.; Fraser, M. P.; Cass, G. R., Air Quality Model Evaluation Data for Organics. 2. C1–C14 Carbonlys in Los Angeles Air. *Environ. Sci. Technol.* **1996,** *30* (9), 2687–2703.
19. Maazawi, M. E.; Finken, A. N.; Nair, A. B.; Grassian, V. H., Adsorption and Photocatalyic Oxidation of Acetone on TiO2: An *in Situ* Transmission FTIR Study. *J. Catal.* **2000,** *191*, 138–146.
20. Griffiths, D. M.; Rochester, C., *Journal of the Chemical Society, Faraday Transactions 1: Physical Chemistry in Condensed Phases* **1978,** (74), 403.
21. Baraka, C.; Gravejat, P.; Guaitella, O.; Thevenet, F.; Rousseau, A., Oxidation of isopropanol and acetone adsorbed on TiO2 under plasma generated ozone flow: Gas phase and adsorbed species monitoring. *Appl. Catal., B* **2014,** *147* (302–313).
22. Gonçalves, H.; Robinet, G.; Barthelat, M.; Lattes, A., Supramolecularity and Photodimerization of Isophorone: FTIR and Molecular Mechanics Studies. *The journal of physical chemistry. A* **1998,** *102*, 1279-1287.
23. Carmona-Quiroga, P. M.; Blanco-Varela, M. T.; Domingo, C.; Martinez-Ramirez, S., Effect of concentration, particle size and the presence of protective coatingsin DRIFT spectra of building materials. *Vib. Spectrosc* **2009,** *50*, 312-318.

24. Meunier, F. C., The power of quantitative kinetic studies of adsorbate reactivity by operando FTIR spectroscopy carried out at chemical potential steady-state. *Catal. Today* **2010,** *155*, 164-171.
25. Pitch, G. M.; Andre, J. S.; Davie, A. J.; Harmon, C. W., Investigations of Water Uptake on Sodium Acetate Trihydrate and Nonhydrate as Model Aerosol Surfaces Using the DRIFTS and ATR Techniques. *Proceedings of the National Conference for Undergraduate Research (NCUR)* **2015**, 300–310.
26. Vogt, R.; Finlayson-Pitts, B., Diffuse Reflectance Infrared Fourier Transform Spectroscopic (DRIFTS) Study of the Surface Reaction of NaCl with Gaseous NO2 and HNO3. *J. Phys. Chem.* **1994,** *98*, 3747–3755.
27. Dilbeck, C. W.; Finlayson-Pitts, B. J., Hydroxyl radical oxidation of phospholipid-coated NaCl particles. *Physical chemistry chemical physics : PCCP* **2013,** *15* (24), 9833-44.
28. Smith, B. C., *Fundamentals of Fourier Transform Infrared Spectroscopy, 2nd Edition*. 2 ed.; CRC Press: 2011.
29. Coronado, J. M.; Katoaka, S.; Isabel, T.; Anderson, M. A., Dynamic phenomena during the photocatalytic oxidation of ethanol and acetone over nanocrystalline TiO2: simultaneous FTIR analysis of gas and surface species. *J. Catal.* **2003,** *219*, 219–230.
30. Loudon, G. M., *Organic Chemsitry* 3rd ed.; The Benjamin/Cummings Publishing Company, Inc.: Redwood City, CA, 1995.
31. Smith, A. L., *The Coblentz Society Desk Book of Infrared Spectra*. 2 ed.; Kirkwood, MO, 1982.
32. Raff, L. M., *Principles of Physical Chemistry*. Prentice Hall: Upper Saddle River, NJ, 2001.

Levine's Marxist Toys

Dean Engle (Humboldt State University/San Francisco State University)

Keywords: Philip Levine, Marxism, poetry, able-ism, William Blake

Perhaps no modern poet can be classified as more American than Philip Levine, who was once described by fellow poet Edward Hirsh as "a large, ironic Whitman of the industrial heartland" (Poetry Foundation). Levine grew up in a Detroit ravaged by the Great Depression and began working in the auto industry at the age of fourteen. These formative experiences never left Levine; themes of industry, class, and the humble laborer struggling against poverty would go on to dominate his work. Through his poetry, Levine attempted to give a voice to the voiceless and was recognized for his achievements when he was named Poet Laureate of the United States in 2011. So, it seems a little ironic that the poetry of this American laureate, this "Whitman of the industrial heartland," features several prominent ideas of Karl Marx—whose communist ideology is often considered antithetical to American capitalism. Levine wrote about poverty and the working class and it always carried a subtle, or sometimes not so subtle, critique of capitalism along with an implied connection to Marxism. Nowhere is this connection to Marxism more prominent than in Levine's poem, "The Toys," which crafts an intense look at an unusual relationship between a boy and a crippled woman. The poem features many questions and ambiguities and uses Marx's ideas of wealth and status to comment on the relationship between children and the physically handicapped in a capitalist economy, as well as exploring humanity's collective commodification within that same capitalist society.

Before delving into the substance of the poem's text, it is important to examine the character concepts Levine is using here and how they relate to both literary tropes and economic theory. That the main conflict of the poem is driven by the character of a child is not incidental. During the industrial revolution, Capitalism expanded as new technology and machinery spread across the Western world. These machines, usually powered from burning coal, required constant upkeep and cleaning. No one was better suited for these treacherous jobs than children, whose small frames and nimble hands allowed them to fit into places larger adults couldn't reach. Of course, due to the dangerous nature of the work, many children were severely maimed or killed. Given these horrifying details it is easy to see how some historians, like J.L. Hammond, "have interpreted child labor in industrial Britain as a crime of capitalism" (Reed). Child chimneysweeps, for instance, became so commonplace, they eventually formed a literary trope common to Romantic and Victorian writers such as Blake and Dickens. A line from Blake's "The Chimney Sweeper" describes the wretched lives of young boys gang-pressed into the industrial age and ends with a scathing indictment of capitalism Marx himself might be proud of, "So if all do their duty, they need not fear harm" (Blake 24). While this harsh life of child labor functioned as a literary trope, a shortcut to highlight the poverty and powerlessness of working-class children, it was also a lived reality for these human beings, one which many of Levine's readers, working class themselves, would likely be aware of. By comparison the boy in Levine's poem is privileged, a child who has likely never had to deal with any of these issues, and that is the point. A working knowledge of child labor and the peril that afflicted generations of working-class children during the Industrial Era, highlights the comparative advantages the boy in the poem has had, while at the same time providing a possible hint at the woman's class background, insinuating that she did not have the same level of privilege as her younger counterpart.

The boy in Levine's poem owns "a selection of his choicest walking toys" (Levine 20). While simply owning toys doesn't automatically mean the boy comes from a wealthy family, the language and the historical context strongly imply it. According to the United States census, the average yearly income for men in 1955, the year the poem was published, was $3,400 (US Census Bureau). Meanwhile the price of walking toys and action figurines ranged, according to the advertising segment of a 1954 copy of Billboard magazine, from around $6.50 on the low end, for a non-moving doll, to $57 on the high end, for a large walking teddy bear (Merchandise). This means that a single walking toy could cost up to nearly 2% of a man's yearly income, a fairly high percentage for a non-essential luxury item. Despite this, the boy in the poem has multiple toys. The phrase "a selection of his choicest walking toys," tells us the boy has enough toys to only show the "choicest" among them. While this does not directly say the boy is wealthy, his ownership of multiple luxury goods, given the price in relation to yearly salary, indicates that his family are members of the bourgeoisie. This puts the boy in an awkward position, one of both privilege and powerlessness. While children's products make up a huge market share of goods bought and sold, it is not the children who are generally buying them but rather the parents (Hill 37-40). Children can ask, beg, plead, or connive to get their wishes, yet ultimately the cash forked over belongs to mommy or daddy. Children are unable to control their lives, locations, or the external forces around them. It is important to understand all these factors to see how Levine characterizes the boy, especially in relation to the crippled woman, who is heavily implied to be in a lower socio-economic class.

Though not directly stated, Philip Levine uses disability as a shorthand, calling upon Western historical and literary connotations to imply the crippled woman is in a lower economic class than the boy. The connotation of disability in canonical Western literature is often of poverty, from biblical lepers too poor to seek treatment to Georgian and Victorian chimney sweeps and coal miners covered in toxic soot. These industrial issues, much as before, were both a literary trope and a lived reality. As discussed above, early industrial machinery could often be fatal or cause serious injury or disability. When the United Kingdom attempted to regulate labor laws in relation to factories and mines in the *Mines and Colliers Act of 1842*, the leader of the opposition, Charles Vane, the Marquess of Londonderry, argued against raising the age for operating dangerous machinery to twenty-one. While he acknowledged this would cause fewer injuries, he did not want to limit the amount of mine operators. "While it afforded a guarantee for security [and] would at the same time not be likely to inflict injury and inconvenience on those engaged in the mines. He believed that fifteen was an age at which there could be no danger in allowing the persons usually engaged to work at those engines" (Mines and Collieries). By today's standards, fifteen is an unacceptable age to put children in charge of heavy duty industrial machinery, and even then it put more workers at risk and, in the long run, caused more work-related accidents for those of lower class, leaving those who were not killed seriously maimed or ill. Levine strongly related to the perils of working-class children thrust into industrial jobs at a young age, as he started working in the Detroit auto industry at fourteen, a full year younger than mine operators a century prior. Though Levine was never seriously injured in the factory, and was able to eventually leave Detroit, Detroit and his experience as a young worker would never leave him, with critic Herbert Leibowitz writing, "Levine has returned again and again in his poems to the lives of factory workers trapped by poverty and the drudgery of the assembly line, which breaks the body and scars the spirit" (Poetry Foundation). Leibowitz does not mince words, specifically drawing attention to the labor conditions of Levine's youth and their effect on both physical and emotional health. Even in our modern day, there is still a genuine link between health and wealth. *The Harvard Gazette*, in a study of American poverty in 2016 noted, "Being poor in the United States is so hazardous to your health... that the average life expectancy of the lowest-income classes in America is now equal to that in Sudan or Pakistan" (Ruell). Many disabled people are caught in a negative feedback loop. Being too sick to work, they are also too poor to afford medical treatment, meaning they become poorer and subsequently sicker (Hudson 37-43). Though none of these literary or historic examples directly tell us that the woman is lower class because of her injury and ill health, like with the child, her role when viewed in the context of these general historical trends and Philip Levine's early life experience, invoke the Marxist concept of powerlessness.

Karl Marx, the leading voice in communist ideology, states in his 1875 "Critique of the Gotha Programme" that "one man is superior to another physically... and supplies more labor in the same time...This equal right is an unequal right for unequal labor. It recognizes no class differences... but it tacitly recognizes unequal individual endowment, and thus productive capacity, as a natural privilege. It is, therefore, a right of inequality" (Marx 1). Levine takes this idea of inequity and marginalization and crafts this poem around two margins, the disabled woman representing poverty and the child representing unearned privilege. Levine purposely juxtaposes these figures together, both symbols of powerlessness from a Marxist viewpoint. Neither can actively make a profit, so neither can be a traditionally central figure. Yet this poem attempts to bring these margins to the center, engaging in a conversation over status.

Marx purposely equated wealth with status. The proletariat were not simply poor; they were fundamentally different from the bourgeoisie. Their difficult lives of work and toil hardened the proletariat as a whole, while, in general, the bourgeoisie had not dealt with the same amount of struggle and therefore did not develop the same conscience towards humanity. For Marx, it was this lack of compassion that made the bourgeoisie problematic, stating "it becomes evident that the bourgeoisie is unfit any longer to be the ruling class in society...because it is incompetent to assure an existence to its slave within his slavery... Society can no longer live under this bourgeoisie; in other words, its existence is no longer compatible with society" (Marx and Engels). To Marx, the bourgeoisie was an evil force, one that deserved a violent overthrow. Levine, however, took these same ideas and softened them, equating the upper class to spoiled children. Like the child with his toys, the bourgeoisie are stuck in a perpetual state of arrested development. Children are not fully developed, either mentally or physically, and cannot be completely responsible for their actions, being too childish to understand their own nature. There is an innocence to the wrongdoing of children, which is stated from the poems opening line.

Levine begins the poem with the concept of forgiveness: "The crippled woman will forgive the boy" (Levine 1). This highlights forgiveness as a major issue throughout the piece, leaving the reader to answer the ambiguous question: What act is the woman forgiving and what implications does this bring to their respective characterizations? In the context of the poem's action, the woman is shown toys by the boy, seeming to mock her disability, "Although she never thought of it that way / A toy was what her motion most resembled" (Levine 3-4). However, seeing as the boy likely represents the bourgeoisie and the woman the working class, this draws an interesting implication. Though the forgiveness can be read as more literal within the poem, as in the woman forgiving the boy for a hurtful question or statement, this forgiveness also carries a greater connotation; a hint that the woman realizes that, just like herself, the boy is randomly born into his position in the world and has no control. He too is powerless. However, this acknowledgment does not erase their differences, as there is still a significant income gap between them. Marx noted that the majority of wealth was owned by a select group of oligarchs, aristocrats, and titans of industry, with the poor forced to live on fragments of that wealth. He even admits that the bourgeoisie do not control the majority of the wealth, referring to the landed gentry, "In England, the capitalist class is usually not even the owner of the land on which the factory stands" (Ryan). This implies the major difference between rich and poor is simply a genetic lottery, a divide between the wealth of work and industry and the wealth of inheritance. While there are some genuine rags to riches stories, such as Andrew Carnegie, most of the world's wealth was inherited, with Thomas Piketty noting in his book, *Capital in the 21st Century*, "it is all but inevitable that inheritance (of fortunes accumulated in the past) predominates over saving (wealth accumulated in the present)" (Buttonwood). It is no one's fault if they are born rich or poor, yet there is still a huge stigma between classes. The poor are deemed poor because they are inferior and have somehow led to their own suffering by not being exceptional enough to turn a profit. This is known as Social Darwinism, that the poor, being inferior, were no use to society and would eventually either die off naturally or be eliminated through eugenicist practices meant to cull those deemed inferior to the upper class on the basis of race or class (Encyclopedia Britannica). This naturally created tension between the bourgeoisie and the proletariat. This tension went both ways, with the poor often resenting the rich and, in some cases, such as in Russia during October 1917, leading violent revolutions against the elite (Smele). This tension is mirrored in the interaction between the woman and the boy. Though they are able to see some common ground, the woman is still viewed as less than the boy. She is viewed as his toy.

The poem equates human beings with machines, specifically toys, suggesting that the lower class exists for the amusement of his or her economic superiors. This concept fits within Marx's theory of alienation, which warns of workers becoming cogs in a machine and losing their humanity. Marx, who lived in the age of industry, was concerned that machinery devalues human labor, "machinery is intended to cheapen commodities and, by shortening the part of the working day in which the worker works for himself, to lengthen the other part, the part he gives to the capitalist for nothing. The machine is a means for producing surplus-value" (Booth). Marx's main fear, in wake of innovations like the assembly line or piecemeal manufacturing, was that workers would lose their skills and become mechanized instruments designed to produce a profit without enriching society. Levine takes these ideas and combines them with the earlier theme of powerlessness, synthesizing it into the idea of being a toy. The woman is referred to as a toy multiple times, "A toy was what her motion most resembled" (Levine 4). This is a comment on her strange movement. Due to her injury, she likely walks in a strange, toy-like way. However, it goes deeper than this. In the second stanza, the speaker states, "This child marveled at her animation / That came within a breath of being human. / The postures of her pain had brought him joy" (Levine 15-17). This implies the boy has a kind of ownership over her, as he finds joy from her pain, the same way a child playing with figurines gains satisfaction when one is injured or defeated. And while the boy can see a base level of humanity, it is not quite enough, it is only "a breath of being human." The ownership of human beings is a very Marxian concept, as Marx believed that by controlling the means of production you controlled the workers as well. Next the boy brings out his other playthings, "A selection of his choicest walking toys" (Levine 20). These cog-based wind-up toys can move, walk, and labor in a way the woman can't. But they, like any good 20th century laborer toiling away at assembly lines, manufacturing machinery piecemeal, lack free will, appearing and acting the same, "Round and round they went, proud and free, / Then stopped with identical smiles on identical features" (Levine 33-34). Here, Marx's feeling of alienation is realized, as a group of people have become so homogenized and similar all they can do is perform their predetermined motions for their superior. However, as Marx dreamed of, there is still always a possibility to overthrow the ruling class.

The final few lines of the poem hint toward a Marxist revolution. Though the crippled woman begins the poem by forgiving the boy, the final lines deal with a rejection of forgiveness and a rise of hostility between both parties. After the toys' display, the woman stares outside and allows her resentment to grow, "The great verbenas, frozen on the trellis, / Boomed her deep uncoiled plea for violence" (Levine 27-28). The transition between forgiveness and violence showcases the discontent of the proletariat over their infantile masters. The infantile master is portrayed more negatively here as well, with the speaker referring to him maliciously twisting the key to keep the toys dancing, "She had no need to speak who knew that sure / And winding hand whose second twist was malice" (Levine 29-30). The child has lost any sympathy for the woman, replacing it with a hardened exterior, which is shown through purposely controlling and manipulating the toys with a cruel intent. The boy is trying to hurt the woman and, by analogy, the proletariat, for his own selfish means. The line, "the postures of her pain had brought him joy" suddenly loses the possible childlike innocence replaced with a cruelty equal to someone intent only on the bottom line, someone comfortable justifying profit as a cause for pain. Perhaps this is a subtle jab at the systemic issues of class, that even well-meaning individuals born into a broken system will abuse it for their own benefit even at the cost of another human being's dignity and livelihood. Either way, the boy is no longer powerless. Gone is any justification due to his age or from the allusions to child-labor; he is now the power of the status quo that needs to be replaced with revolution.

Levine's use of literary tropes and Marxist philosophy to highlight the plight of the working class comes as no surprise. Published in 1955, this is an early Levine piece and relies on a more traditional format, borrowing aspects from other modernist poets (Stacey). In his later career he would break this mold, defining his own style but always keeping his original working-class background at heart. Due to this background, there is a sympathy given to the crippled woman, from her capacity to forgive to her need for violence. Through these contradictions she is portrayed very much as human, even as she is compared to a toy. Levine's poem critiques the system surrounding people, but not the people themselves. It is not that they are guiltless, but rather that they are the products of something greater than themselves. The beginning of the second stanza echoes this claim. Though it seems out of place at first, the line "Was it the onus of a father's crime / That had yoked her tight within a cage of tears?" is given meaning when considered in this light (Levine 11-12). The father's crime doesn't refer to the woman's father or the boy's, but rather their collective forefathers, the bankers and traders and merchants who built capitalism up and turned it into an invisible force, the unseen hand of the market throttling our throats. To Levine, we are all children of this system, powerless to stop it. Some of us may be born healthy and wealthy, others poor and hurt, yet we are all part of it, dancing along because we must, the keys on our backs winding and turning like clockwork, but still trying to remain human along the way.

Works Cited

Blake, William. "The Chimney Sweeper: When My Mother Died I Was Very Young by William Blake." *Songs of Innocence* , William Blake, 1789,www.poetryfoundation.org/poems/43654/the-chimney-sweeper-when-my-mother-died-i-was-very-young.

Booth, Adam. "Marx's Capital: Chapters 15 - The Machine." *Socialist Appeal*, Socialist Appeal, 12 Aug. 2014, www.socialist.net/marx-s-capital-chapters-15-the-machine.htm.

Buttonwood. "Inherited Wealth." *The Economist*, The Economist Newspaper, 18 Mar. 2014, www.economist.com/blogs/buttonwood/2014/03/inequality.

Encyclopædia Britannica. "Social Darwinism." *Encyclopædia Britannica*, Encyclopædia Britannica, Inc., 27 July 2017,www.britannica.com/topic/social-Darwinism.

Hill, Malcom. "Children's Role in the Domestic Economy." *Journal of Consumer Studies*

& Home Economics, Blackwell Publishing Ltd, 26 July 2007,onlinelibrary.wiley.com/doi/10.1111/j.1470-6431.1992.tb00497.x/pdf.

Hudson, Christopher G. "The Social Class and Mental Illness Correlation: Implications of the Research for Policy and Practice," *The Journal of Sociology & Social Welfare*: Vol. 15 : Iss. 1 Article 3. 1988, http://scholarworks.wmich.edu/jssw/vol15/iss1/3

Levine, Philip. "The Toys." *Poetry* , vol. 86, 1955, pp. 78–79, www.poetryfoundation.org/poetrymagazine/browse?contentId=26672.

Marx, Karl, and Friedrich Engels. *The Communist Manifesto*. 61st version, 1848, *Project Gutenburg*, www.gutenberg.org/cache/epub/61/pg61.html.

Marx, Karl. "Critique of the Gotha Programme." *Critique of the Gotha Programme*, vol. 3, Die Neue Zeit, 1875, pp. 18–30,www.marxists.org/archive/marx/works/1875/gotha/ch01.htm.

"MINES AND COLLIERIES." *MINES AND COLLIERIES. (Hansard, 14 July 1842)*, Parliament.UK, api.parliament.uk/historic-hansard/lords/1842/jul/14/mines-and-collieries#s3v0065p0_18420714_hol_51.

"Merchandise." *The Billboard: Nov 6, 1954*, Nielson Business Media, Inc, 1954, pp. 69–73, books.google.com/books?id=myEEAAAAMBAJ&pg=PA73&lpg=PA73&dq=walking+toys+1955+price&source=bl&ots=B35c5yGCjx&sig=ACfU3U2e5ddoL1RsEtry2TqFtRupuBEQvA&hl=en&sa=X&ved=2ahUKEwibkJarkr7pAhUHZKwKHflDCI04ChDoATAFegQICh-AB#v=onepage&q&f=false.

Poetry Foundation . "Philip Levine." *Poetry Foundation*, Poetry Foundation, www.poetryfoundation.org/poets/philip-levine.

Reed, Lawrence W. "Child Labor and the British Industrial Revolution." *Mackinac Center*, Mackinac Center for Public Policy, 7 Dec. 2001, www.mackinac.org/3879.

Reuell, Peter. "For Life Expectancy, Money Matters." *Harvard Gazette*, Harvard University, 11 Apr. 2016, news.harvard.edu/gazette/story/2016/04/for-life-expectancy-money-matters/.

Ryan, Alan. *On Marx: Revolutionary and Utopian*. 1st ed., Liveright, 2014, books.google.com/books/in_england_the_capitalist_class_is_not_even_the_owner&f=false.

Smele, Dr Jonathan. "History - World Wars: War and Revolution in Russia 1914 - 1921."*BBC*, BBC, 10 Mar. 2011, www.bbc.co.uk/history/worldwars/wwone/eastern_front_01.shtml.

Stacey, David. "English 120." 7 Dec. 2017, Arcata, Ca, Founder's Hall.

US Census Bureau. "Income of Persons in the United States: 1955." *The United States Census Bureau*, census.gov, 4 Sept. 2018, www.census.gov/library/publications/1956/demo/p60-023.html.

Reclaiming Indigenous Women's Roles in the 21st Century

Crystal Miller (Humboldt State University/ Arizona State University)

Keywords: Native, identity, Native women, traditional, anti-colonial, invasion, colonial consequences, healing

Terminology

In this article, the term *Indigenous* and *Native* (with a capital "N") will be used interchangeably to describe the original inhabitants of an area. The term *invasion* will describe the invasion of Indigenous land by Europeans. *Self-Identity* or *Identity* will be used to describe the traditional idea of oneself pre-invasion. The phrase *White World* is not to be described as a different space or world, it will be used to describe those who believe that all ideas and life should fit into a Western worldview, or come from it, and that those who function in it are superior.

Decision making and leadership roles for Native women have shifted disproportionately because of circumstances that stem directly from colonization. Although Native women have been resourceful in how they have actively worked to rebalance themselves as leaders in traditional matrilineal societies and reclaim their respective statuses in their communities, one way in particular that Native women have reclaimed their identity as knowledge holders and leaders is by navigating the academy successfully and returning home to their nations, or doing work that has a direct impact on their nations to help create change. A shift in balance in contemporary Native nations has caused Native women to be oppressed and marginalized from both Native communities and outside communities as a direct result of colonization. Traditional Native societies once held the women at very high standards and recognized that although their roles in their community differ from a man's role, they were still held in equal value to the man: everyone understood their roles and they all mattered in the community (Green). Invasion of the land by settlers began a darkness that has haunted, and still does, Native people since contact and has birthed many past and present issues that affect their communities today. Historical and intergenerational trauma, brought on by assimilation, ethnocide, forced removal, and on-going oppression, have been the root of this shift. Colonization has forced trauma and lateral oppression on current Native communities, making it difficult to strive, along with forcing cultural loss and self-identity issues to take place (Garza). These losses have reinforced the shift in balance regarding roles and respect for Native women. In addition, colonization has created many trickle effects of social, emotional, and spiritual grief. However, despite the long history of colonization and systemic, ongoing oppression, Native women have worked endlessly to break down barriers, reclaim their identity and traditional knowledge, and define their success on their own terms through academia, in order to break generational curses to revitalize and heal their nations.

To understand the effects of colonization, or "the formal and informal methods, (i.e., behaviors, ideologies, institutions, policies, and economies) that maintain the subjugation or exploitation of Indigenous Peoples, lands, and resources (Wilson & Yellow Bird, 2005)" (Garza, 2015)", on Native women and gender roles, it is important to understand that colonization has plagued Native nations since the invasion of settlers and has yet to be extinguished. It has stripped people of their humanity and lives with its willful force to kill and remove anything in its way. It has destroyed Native families, cultures, traditions, and land throughout the 527 years post-invasion. U.S. Colonization is still present and works to keep Native populations oppressed and the policies and laws implemented by the U.S. have played a major role in colonization. Natives were subjected to reservations in the late 19th century, which forcibly removed Native nations to confined territories in which they were to stay bound to ("The Reservation System | Native Americans"). This made it difficult for Native people to steward the land and continue their traditional lifestyle of worship, hunting, and living freely. The first official off-reservation boarding school began in 1879 and was known as the Carlisle Indian School ("History and Culture"). The boarding school system was designed to assimilate Native children into the "white world" and Christianity, which included white language and culture and was a contribution to the loss of Native identity and ethnocide ("History and Culture"). At these boarding schools, children who were stripped from their families and forced to go were almost always beaten, raped, or starved if they were caught practicing their traditional way of life and/or language ("History and Culture"). The experiences that came with the attendance of boarding schools were almost always violent, and, naturally, the violence was internalized by the attendees and unintentionally passed down through generations. The Dawes Act of 1887 authorized the U.S. government to break up land into checkerboard land assignments and allowed individual lots to be parceled out ("Dawes Act", 1887). This act weakened the kinship and force of Native nations by dismantling them through isolation, reinforcing their "divide and conquer" systems ("Dawes Act", 1887). In 1956, the Indian Relocation Act was a part of the Indian termination policy that encouraged Native people to leave reservations and assimilate into the "white" world (Library). All of these policies caused emotional and spiritual grievances for Native people broadly, and Native women specifically, because they influenced major cultural loss. Many times, if not always, Native people were backed into a wall to comply with these policies because their families or survival was threatened. Because of these policies, Native nations who unintentionally internalized the trauma began to adopt Western worldviews as a means of survival and with that internalization and adoption arose accepted violence. Through ethnocide, forced assimilation, and the internalization of Native people's experiences passed down intergenerationally, a loss of cultural identity occurred that has led to contemporary social problems and the shift of balance in gender roles in Native society.

"The concept of historical trauma, also called intergenerational trauma…involves exposure of an earlier generation to a traumatic event that continues to affect the subsequent generations" (Cole 2006). These internalized experiences of harm and trauma have been an invisible aspect that contributes a great deal to the shift of balance in gender roles in Native societies. The major historical policies discussed in the previous section have influenced the shift of balance of equality, reaffirmed the dominance of the Eurocentric society, and began the internalization and projection of violence, from both Native and non-Native people, as the societies shifted. Winona LaDuke, an enrolled member of the White Earth Mississippi Band of Ojibwe (Bear Clan) and well-respected environmentalist, economist, and writer, who is most known for her work on tribal land claims & environmental justice, states:

> We collectively find that we are often in the role of the prey to a predator society whether through sexual discrimination, exploitation,

> sterilization, absence of control over our bodies, or being the subjects of repressive laws and legislation in which we have no voice. This occurs on an individual level, but equally and more significantly on a societal level. It is also critical to point out at this time most matrilineal societies, societies in which governance and decision making are largely controlled by women, have been obliterated from the face of the Earth by colonialism and industrialism. The only matrilineal societies that still exist in the world today are those of indigenous nations. Yet we also face obliteration" (LaDuke, p. 213)

As Winona expresses, colonialism, in its nature, has been destructive to Native nations and, because of the traumatic events from colonization, many societal ills have surfaced that have made success for Indigenous women and restoration of traditional societies extremely hard. The societal ills in Native communities that emerged from colonization are contemporary problems that have influenced a society that citizens of that society attempt to fix. It is, most of the time, a consequence of factors beyond their control. In an article written by Lisa M. Poupart, she states that, "Virtually nonexistent in traditional tribal communities prior to European invasion, contemporary American Indian communities struggle with devastating social ills including alcoholism, family violence, incest, sexual assault, fetal-alcohol syndrome, homicide, and suicide at startling rates similar and sometimes exceeding those of white society" (Poupart, p. 88). Societal ills and statistics, according to the American Psychological Association, include Native people having the highest alcohol consumption on a weekly basic (Cole 2006). A connection has been made between alcohol abuse in Native nations and low self-esteem, loss of cultural identity, lack of positive role models, history of abuse and neglect, self-medication due to feelings of hopelessness, and loss of family and tribal connections (Brown-Rice)—all of which are a direct result of colonization. The problems that are found in contemporary Native nations have been found to be tied to past trauma and oppression that these nations have endured generationally. Ultimately, birthing children into a destructive cycle with no adequate form or method of healing, unintentionally continues the cycle. The societal ills that are present in contemporary Native nations are contributing factors and a piece of the history that has unintentionally allowed the shift in balance between these societies.

Violence is something that Native people have been historically exposed to through many different avenues and tactics. Through historical trauma, violence was internalized by Native populations as a collective and intergenerationally passed down; contributing to the shift of balance in Native gender roles. Traditional ways of life had begun to shift, balance in Native nations was beginning to whither, and, more times than not, the women of these nations were the most affected. On some reservations, Native women are murdered at ten times the national average (Walker). Colonial consequences have affected the women because it has allowed them to be objectified in a number of ways. Modern society has seen Native women as sex symbols with no regard to real consequences when violence is/has been projected on/towards them. Disney, who used one of the first documented cases of human trafficking on Native women and romanticized it, is a prime example. They made Indigenous women a sexual fantasy by the production of the movie Pocahontas. 84% of Native women have experienced violence in their lifetime. Domestic violence and physical assault are estimated to be as much as 50% higher in Native communities than the next most victimized demographic (Formerly Family Violence Prevention Fund) (Rosay). More than 1 in 2 women have experienced sexual assault (Walker). This is a devastating result of the balance shift in Native nations because it has forced the women of these nations to be seen as less than, in both their communities and in the "white" world.

Statistics have defined the scale of the problem but there is more to be done to convey the epidemic proportions of the issue and the experiences that come with it ("Safe Women, Strong Nations"). The statistics tell a small portion of the story and fail to account for the impacts the survivors have gone through ("Safe Women, Strong Nations"). The Indian Law Resource Center's statistics show that Native children who are exposed to violence suffer rates of PTSD three times higher than the rest of the general population in America ("Safe Women, Strong Nations"). These statistics provide us with insight on how violence against Indigenous women has a direct impact on the next generation of Native children, being the portal that children come into this world through. If not addressed and corrected, it will only continue. The Indian Law Resource Center uses their platform as a project that raises awareness to get federal attention and support to end violence against Native women. Violence has been normalized by Native communities and the uncontrollable acceptance of this such behavior has set the tone for the exploitation and abuse of Native women, majorly at the hands of non-Native men, that continues into the 21st century due to patriarchy and white supremacy brought on by the colonizers.

Due to a long history of assimilation, the family dynamics in these nations began to change and the patriarchal systems that European invaders functioned under began to influence Native nations. When speaking on the leadership roles in the Navajo Nation, Jennifer Nez Denetdale, citizen of the Navajo Nation, professor of American Studies at the University of New Mexico, and the first-ever Dine/Navajo to earn a Ph.D. in history, states, "Although written reports do not mention women as leaders or chiefs, Navajo oral tradition and other accounts make note that it was not unheard of for women to serve as headmen or chiefs. Further, early American accounts have noted Navajo women's presence in council proceedings between Navajo and American leaders… they influenced the decisions that male leaders made on behalf of the people" (Nez Denetdale, p. 177). Prior to European arrival, women in Native nations had an influence in decision making and were well-respected as leaders. Due to European influence over time, gender roles in Native nations had completely changed. Jennifer Nez Denetdale later mentions in her same chapter, when describing what changed in the gender roles and why, that, "Navajos were subjected to government rule based on Western democratic principles. This model looked to Navajo men to fill leadership roles. As the anthropologist Christine Conte and the Navajo attorney Genevieve Chato note, this Western form of government is patriarchal and has undermined Navajo women's traditional rights, including land-use rights, property and livestock rights, and the right to primary care and control of their children… Like white American women, Navajo women were expected to relegate themselves to the domestic realm, which is associated with little political or economic power" (Nez Denetdale, p. 179). This new expectancy of Native women, being similar to American women, was manifested by the ideas that Native communities adopted from Western worldviews. Although the shift of balance from traditional societies to modern (Western) societies has happened, contemporary Native women have been actively working to recognize past and current issues and find ways to heal them appropriately and restore balance in their communities.

Indigenous women have proven that there are many different ways they have successfully been able to resist this shift in balance and have begun working to restore the equality of roles in their communities. Though there are a multitude of ways that Native women have been able to do this, one way is how Native women have navigated through academia successfully and used their knowledge to help re-balance their traditional communities and leadership roles, but even accomplishing this it has been difficult. In academia, white academics have attempted to tell the story of Native people, the issues and ills within their own communities and themselves, which has made it harder for Native women to create their own success within academia. They are told how to co-exist in academia with white scholars from white scholars and that the only way to be successful is to be successful in the eyes of white scholars but, here too, Native women have resisted. They began to define their own success. They were not attending post-secondary institutions with the intent to obtain a high paying job or status in Western eyes; they were doing it as a tool to combat contemporary issues that their communities are suffering from as a direct result of colonization. It is not the degree, the paper, or the standing from Western constructs that give these Native women their identities and power back. It is what they do with their success for their people, their communities, and the culture that is giving them their power and voices back in both the Indigenous and Eurocentric societies. Native women have opened up doors to address many different issues that have been ongoing in Native nations and have given Native people a voice on a larger platform than ever before. As Vikki Eagle-Bear, Sicangu/Oglala Lakota citizen of the Rosebud Sioux Tribe and Doctoral student in Education Administration at the University of South Dakota, states:

> Across the world, Indigenous women are mobilizing for our rights to protection and equality, more so, our right to a seat at the table where major decisions are being made that impact our lives. I represent a nation of people who come from a matriarchal society. Unfortunately, as a result of exposures to generational trauma and colonization the Lakota succumbed, culturally, until the past few decades. Those who continued practicing the Lakota ways and those who reclaimed their ways, know that we live on a female planet and all that is created by the female is the most powerful. Armed with this knowledge, I know my role as a Lakota woman at home, in my community, and in the world. In my community, of those who have obtained post-secondary degrees, all are women. At the tribal level, majority of programs are led by women and fifty percent of our Tribal Leadership (Council) is women (V.Eagle-Bear 2019).

Women across Indian country, especially those in Vikki's community, are allowing issues to be brought to the forefront and are the force that has been needed in Indian country. They have re-positioned themselves as leaders and decisionmakers in their communities by finding a middle ground between what they know in their community and what they learned in the academy to be successful in their nations. By doing this, they have been able to dismantle Western ideologies of how Native nations should function in the 21st century. Not only are Indigenous women reclaiming their identity as powerful beings, they are changing ideas of Native nations in the white world, forcing the powerful in the white world to listen, forcing them to acknowledge what they work so hard to forget, and forcing Western education to include Native history from Native worldviews.

Academia has allowed many Native women the platform to help create change through Indigenous research and scholarly work. It has paved the way for those to come next and has pushed Western education to accept and acknowledge Native people's presence on a parallelism. Women like Maria Yellow Horse Brave Heart, Winona LaDuke, Elizabeth Cook-Lynn, Debra Holland, and Sharice Davis are only a small fraction of successful Indigenous women on the long list of women who have been able to successfully use academia to accomplish their ultimate goals; restoring Native nations. They have created pathways for young scholars to follow and have used their success in academia to highlight issues that stem from colonialism. Elizabeth Cook-Lynn has used her platform to protect tribal sovereignty and establish that Native nations within the boundaries of the U.S. have yet to reach a level of postcolonialism because colonialism still has a major influence on Native people. Her success in the academy has allowed her to protect Native cultures and communities by positioning herself on an equal level as non-Natives in academia and to question ideas that come from a Western worldview of Native people. Maria Yellow Horse Brave Heart has dedicated her work and research on understanding historical trauma, and how it affects Native nations, and has opened the door to endless possibilities of exposing colonialism as a weapon. Winona LaDuke, a longtime environmentalist activist for her community White Earth, has challenged colonial systems in and out of the court room. Debra Holland and Sharice Davis have a voice in congress that allow Native people's issues to be heard, rather than further pushed to the bottom of the list. The work that all of these specific scholars have done, forces Western education to not only acknowledge its wrongs but also acknowledge the outcomes of their wrongs and how it continues to affect Native people. Because it is a process to dismantle colonialism and all of its negative effects in Native nations, persistence is the key to drastically force change in the Western world that will positively benefit Native nations by helping to revitalize traditional societies and restore the balance in these communities. Alexandra Mojado, Cherokee Nation/Pala Band of Mission Indians/Luiseno/Ute/Paiute/Sho-ban and J.D. holder from University of Arizona, James E. Rogers College of Law, states, "In my job, I often look first to Tribal law, cases, and culture when making arguments instead of relying on state or federal law and cases" (A. Mojado, 2019). This style of practice describes how the outcomes of Federal Indian Law and cases can change when ideas stem from Indigenous worldviews. This is important because it allows a decision to be made from an Indigenous perspective to serve an Indigenous individual or community and to provide an Indigenous decision—decolonizing the court system. Having Native lawyers in positions who are capable of separating Western ideas of law from traditional thinking is something that helps maintain not only identity but sovereignty as well, and Alexandra's success in academia has helped her reclaim her place in a traditional society as a leader. Native women have been actively working, in an effort of resistance, to position themselves to create change and ultimately bring extended knowledge back to their communities to help break traumatic cycles.

Indigenous women are taking a stand and resisting by working to break generational curses. They are revitalizing cultures and strengthening their voices as proud Native people. They are continually forcing the dominant society to be uncomfortable with their demanding presence and recognition of who they are. They are no longer standing in the shadows and fitting the mold that the invaders so greatly forced them to fit in. They are working to beat negative statistics and stereotypes that they are born into and repositioning themselves as the leaders they were always supposed to be. The significance of Native women succeeding in academia and life in general, is that they are almost always giving back to their communities. Success in the Western world does not define them and they do not allow that to be the standing from which they demand respect. What Native women do *with* their success is what is defining who they are and ultimately defining their success. The route in which Native women are taking to not only reclaim their identity but also change history by doing so, is a reflection of their knowledge of the past, the present, and their visions for the future. With all odds against Native nations broadly, and Native women specifically, perseverance and hope still stands. Resistance has been the foundation that has fueled Native women to work endlessly to help rebalance themselves as leaders in matrilineal societies.

Bibliography

Abadian, S., & Sen, Amartya. (1999). From Wasteland to Homeland: Trauma and the Renewal of Indigenous Peoples and Their Communities, ProQuest Dissertations and Theses.

Brown-Rice, Kathleen. "Examining the Theory of Historical Trauma Among Native Americans." *The Professional Counselor*, The Professional Counselor, 10 July 2020, tpcjournal.nbcc.org/examining-the-theory-of-historical-trauma-among-native-americans/.

Campbell, Alexia Fernández. "How America's Past Shapes Native Americans' Present." The-Atlantic, Atlantic Media Company, 13 Oct. 2016, https://www.theatlantic.com/business/archive/2016/10/native-americans minneapolis/503441/.

Cole, N. (2006). Trauma and the American Indian. In T. Witko (Ed.), Mental health care for urban Indians: Clinical insights from Native practitioners (pp. 115-130). Washington, DC: American Psychological Association.

"Dawes Act (1887)." Our Documents - Dawes Act (1887), www.ourdocuments.gov/doc.php?flash=true&doc=50

Deloria, Vine. Custer Died for Your Sins. 1970.

Eagle Bear, Vikki. Personal Interview. 19 Jul. 2019.

Evans-Campbell, T. (2008). Historical trauma in American Indian/Native Alaska communities: A multilevel framework for exploring impacts on individuals, families, and communities. Journal of Interpersonal Violence, 23(3), 316-338.

Evans-Campbell, T., Walters, K. L., Pearson, C. R., & Campbell, C. D. (2012, September).

Indian Boarding School Experience, Substance Use, and Mental Health among Urban Two-Spirit American Indian/Alaska Natives. Retrieved October 22, 2018, from https://www.ncbi.nlm.nih.gov/pmc/articles/PMC5446670/

Garza, J. (2015). The impact of historical trauma on Native Americans.

Green, Rayna. Women in American Indian society. New York, NY: Chelsea House Publishers, 1992

Formerly Family Violence Prevention Fund. "The Facts on Violence Against American Indian/Alaska Native Women." Futures Without Violence, https://www.futureswithoutviolence.org/userfiles/file/Violence%20Against%20AI%20_N%20Women%20Fact%20Sheet.pdf. Accessed 19 Nov. 2017.

Hinton, D., & Good, B. (2016). *Culture and PTSD: Trauma in global and historical perspective* (Ethnography of political violence). Philadelphia: University of Pennsylvania Press.

"History and Culture." Native American History and Culture: Boarding Schools – American Indian Relief Council Is Now Northern Plains Reservation Aid, www.nativepartnership.org/site/PageServer?pagename=airc_hist_boardingschools

LaDuke, Winona. The Winona LaDuke *Reader : a Collection of Essential Writings*. Stillwater, MN :Voyageur Press, 2002.

Library, World. "*Indian Relocation Act Of 1956.*" *Indian Relocation Act Of 1956* | World Library –EBooks | Read EBooks Online, www.worldlibrary.org/articles/indian_relocation_act_of_1956.

Mojado, Alexandra. Personal Interview. 17 Jul. 2019.

Nez Denetdale, Jennifer. "Chairmen, Presidents, and Princesses: The Navajo Nation, Gender, and the Politics of Tradition." *Native Historians Write Back: Decolonizing American Indian History*, by Susan A. Miller and James Riding In, Texas Tech University Press, Lubbock, TX, 2011, pp. 176–186.

Poupart, L. (2003). The familiar face of genocide: Internalized oppression among American Indians. Hypatia, 18(2), 86.

Ringel, S., & Brandell, J. (2012). Trauma: Contemporary directions in theory, practice, and research. London: SAGE.

Rosay, André B. "Violence Against American Indian and Alaska Native Women and Men." *National Institute of Justice*, National Institute of Justice, 1 June 2016, nij.ojp.gov/topics/articles/violence-against-american-indian-and-alaska-native-women-and-men.

Safe Women, Strong Nations. (n.d.). *Indian Law Resource Center*. https://indianlaw.org/safewomen

Sue, D.W., & Sue, D. (2012). *Counseling the culturally diverse: Theory and practice.* (6th ed.). New York, NY: John Wiley & Sons, Inc.

"The Reservation System | Native Americans (Article)." *Khan Academy*, Khan Academy, www.khanacademy.org/humanities/us-history/the-gilded-age/american-west/a/the-reservation-system.

Walker, Jana L. "*Ending Violence Against Native Women*." Ending Violence Against Women I Indian Law Resource Center, Indian Law Resource Center, http://indianlaw.org/issue/ending-violence-against-native-women. Accessed 19 Nov. 2017.

Whitbeck, L., Adams, G., Hoyt, D., & Chen, X. (2004). *Conceptualizing and measuring Historical trauma among American Indian people.* American Journal of Community Psychology, 33(3/4), 119-130.

Whelshula, M., & Colorado, Apela. (1999). Healing through Decolonization: A Study in theDeconstruction of the Western Scientific Paradigm and the Process of Retribalizing Among Native Americans, ProQuest Dissertations and Theses.

Who Was Pocahontas? 1 Oct. 1970, www.niwrc.org/who-was-pocahontas.

Woolford, Benvenuto, Hinton, Fontaine, Woolford, Andrew John, Benvenuto, Jeff, . . . Fontaine,-Theodore. (2014). *Colonial genocide in indigenous North America.*

Yellow Horse Brave Heart, M. (n.d.). *Historical Trauma and Unresolved Grief: Implications for Clinical Research and Practice with Indigenous Peoples of the Americas* [Historical Trauma]. Retrieved October 28, 2018, from https://www.ihs.gov/telebehavioral/includes/themes/newihstheme/display_objects/d cuments/slides/historicaltrauma/historicaltraumaintro_011113.pdf

Scaling Resistance with Channel Length for Carbon Nanotube Networks

Cade Freels (Humboldt State University), Carla Patricia Quintero (Humboldt State University), Samantha A. Anger (Humboldt State University), Will Gannett (Humboldt State University), Ruth Saunders (Humboldt State University)

Keywords: physics, carbon nanotube, transistor, biosensor

Introduction

Single walled carbon nanotubes (CNTs), which are an individual rolled sheet of graphene, have been the focus of extensive research since their discovery in 1991 [1]. Single walled carbon nanotubes are attractive because of their mechanical, electrical, thermal and lightweight properties [2]. One promising electrical feature of these tubes is that, depending on the angle at which the sheet is rolled up, the tube can be either metallic or semiconducting, allowing them to find numerous industrial applications [3]. The building of well-structured carbon nanotube systems is not yet practical, as the ability to position CNTs and reproducibly control the dimensions and chirality of the nanotube is limited. However, devices that make use of CNTs through the creation of CNT networks, in which no real control of the position is needed and effects due to individual variations are reduced, are possible to manufacture [4]. These thin films of CNTs have been studied as thin-film high performance field effect transistors [5] and biosensors [6]. The devices consist of a thin film of randomly distributed metallic and semiconducting CNTs placed in between a source and drain electrode as illustrated in Figure 1.

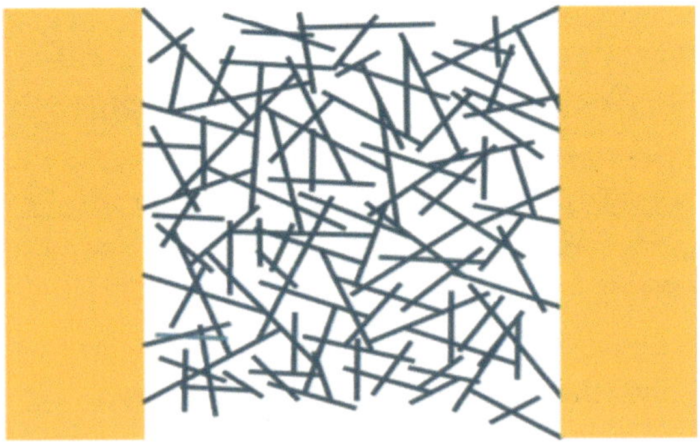

Figure 1: Schematic showing source and drain electrode with randomly distributed carbon nanotubes between them.

The electronic behavior of the network is dependent on geometric factors such as the length of the channel between the source and the drain, and the length and density of the carbon nanotubes. It also depends on the properties of the wires, such as whether they are metallic or semiconducting, or whether they have formed bundles. The ability of the network to act as a transistor is dependent on the presence of semiconducting wires. The length scale of the CNTs are shorter than the mean free path of electrons, therefore the source of resistance of the network is predominantly due to the junctions between wires [7, 8]. The resistance of the junction between a metallic and a semiconducting CNT has been measured to be two orders of magnitude higher than a junction between two metallic CNTs [9]. The metallic-metallic junctions also do not respond to the presence of an external electric field, making the transistor performance less effective. The junction between a metallic and semiconducting CNT can be modeled as a Schottky barrier. The junctions between metallic CNTs will not respond so their presence alters the on-off current ratio, limiting the transistor performance. Originally, production of CNTs produced ~⅓ metallic wires. Having such a high proportion of metallic CNTs increased the presence of metallic junctions and even the presence of fully percolating metallic pathways connecting the source and the drain. Improvement in production methods [10] have reduced the number of metallic CNTs, giving 99 % semiconducting wires. These improvements make CNT networks an exciting option to be used as transistors, specifically ones used in cell phones and other small everyday devices. Further progress of CNT networks may allow them to play a central role in the future advancement of electronic devices.

Given the variety of proposed applications for CNT networks, creating a working accurate model to explore the behaviors of these networks is crucial to creating reproducible and reliable devices. Our group's computational model of CNT networks adds to other models previously considered, and allows us to place our results in the context of previous computational findings to ensure that our model is accurately characterizing the behavior of CNT networks. These models, ours included, allow researchers to generate CNT networks with a random assortment of carbon nanotubes, as well as providing the ability to numerically adjust certain variables within the network. The variables most commonly adjusted in such models are channel length [11, 12], density of nanowires [13], and percentage of semiconducting nanowires present [5]. Changing these variables, as well as changing the presence of charged molecules, produces a change to the measured resistance of these networks. Different changes to the network allows for helpful information depending on the network's intended use.

Similar to the models used in the research above, we have been utilizing a computational model for carbon nanotube networks to investigate them and determine how these networks will function in applications. One component of this computer model grants us insight into the electrical properties of these networks. Specifically, we have been investigating how the resistance of these networks scales as the channel length of the network varies for differing concentrations of metallic tubes within the network. This allows us to better inform on the usefulness of carbon nanotube networks as field effect transistors and can provide novel approaches to the field of medicine through their use as biosensors.

Methods

The following is our computational algorithm to model the behavior of a network of carbon nanotubes distributed between two electrodes.

The first stage is the generation of the two-dimensional network of randomly distributed wires. The area of the network is defined by the channel length, L_c and channel width, W_c. A random initial point (x_i, y_i) is generated within that area. An angle θ is generated assuming a uniform distribution and is used with a fixed wire length, L_s to find the final point of the wire (x_f, y_f). This process is repeated, populating the network until the required density is reached. Each generated wire is assigned a chirality of either metallic or semiconducting based on an entered probability. The electrodes are defined on either end of the channel length.

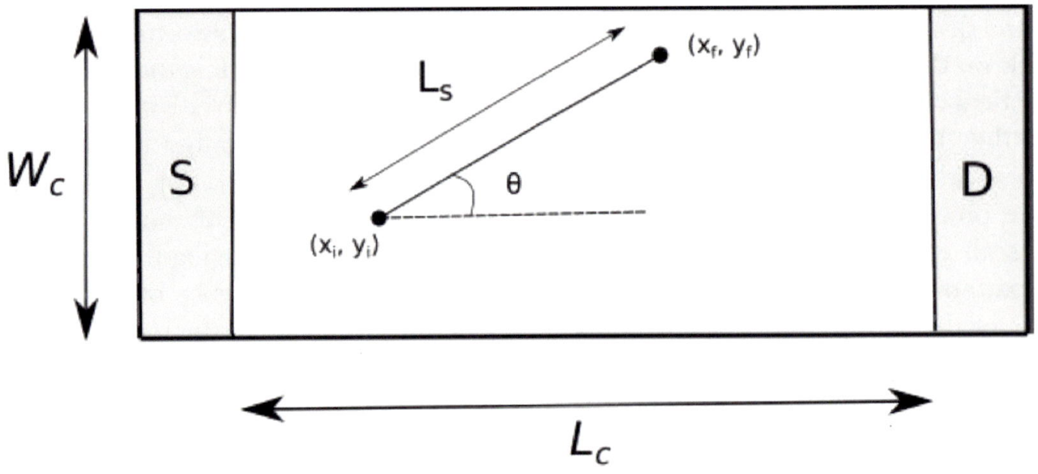

Figure 2: Diagram of how our computer model generates carbon nanotubes.

An example of a diagram showing these parameters is shown in Figure 2.

The next stage is to map the junctions within the network. Each pair of the wires needs to be evaluated. To save on time, a sweep line algorithm is used to match only pairs that are close together. Using the equations of the lines that the wires lie along, the intersection point is found. If the intersection point occurs within the segment of the wire then we record that as a junction.

This information is stored in an adjacency matrix. Here we are making use of some concepts from graph theory, where a network is described as a series of vertices connected by edges of a certain weight. An adjacency matrix can be used to store information about connections between vertices in graph theory. In the generated network the interesting element is the junctions between wires. Each wire is considered to be a vertex and the junction is the edge that connects them. The adjacency matrix is an n x n matrix, with elements Aij being zero if there is no connection between wires i and j and

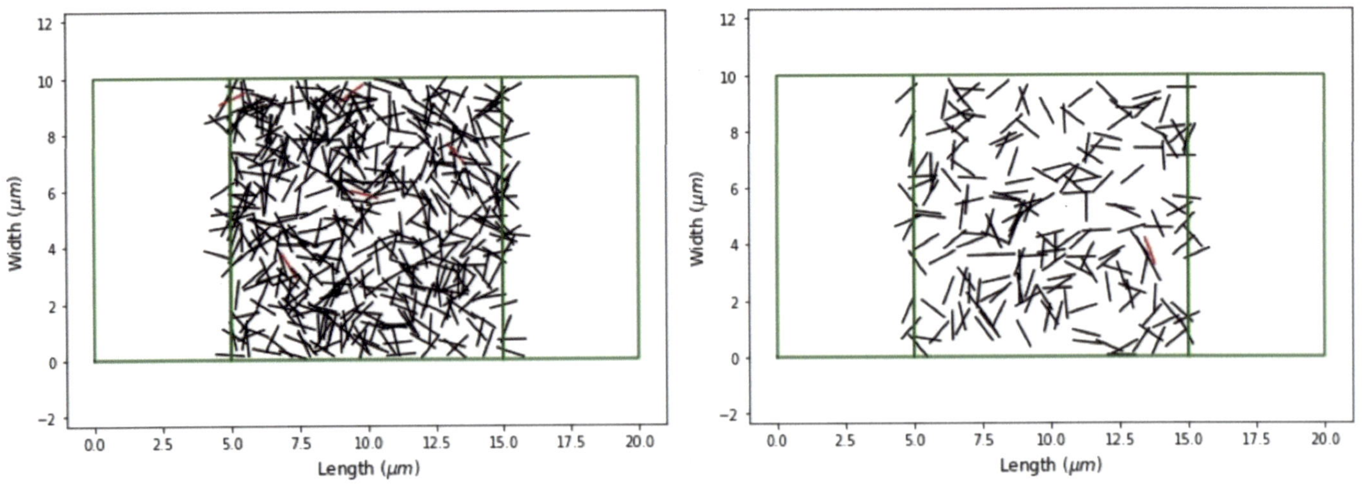

Figure 3: Networks of different tube density showing the wires generated by our model for a channel length and width of 10 μm, tube length of 1 μm, and 1% metallic wires (shown in red). The image on the left corresponds to a tube density of 5 tubes / μm^2 , while the image on the right corresponds to a tube density of 2 tubes / μm^2.

is non-zero if there is a connection. The diagonal elements of the matrix are zero. The value of A_{ij} for connecting wires is the weight of the edge, or in this case, the conductance of the junction, which is given by 1/R where R is the resistance of the junction. Our model specifically deals with three different junction types. There are three junction types depending on the chirality of the connecting wires, metallic-metallic (m-m), semiconducting-semiconducting (sc-sc) and metallic-semiconducting; the value of R will depend on which types are in the connecting pair. We can use the adjacency matrix to calculate the effective resistance of the entire network. Examples of generated networks with different densities are shown in Figure 3.

We can determine what the resistance of the network will be if we know the channel length of the network. From percolation theory, the relationship between these two is established as

$$R \ \alpha \ \left(\frac{Lc}{Ls} \right)^m \tag{1}$$

where m is a constant. By taking the logarithm of both sides to get a linear relationship, this equation becomes

$$Log\,R = m\,Log\,Lc + C \tag{2}$$

where m is a constant scaling factor between the resistance of the network and the channel length of the network, and C is the logarithm of the tube length, a constant. A log-log plot of the resistance versus channel lengths, for networks with a density of 1.6 $tubes/\mu m^2$, and various amounts of metallic wires present in the networks, was produced. We determined our metallic wire probability within the networks from previous industry standards (approximately 30% metallic wires), as well as current yields (1% metallic). In addition, we hoped to further validate our model by comparing our simulation results to experimental results that yielded carbon nanotube networks with yields of approximately 33% metallic wires [3]. Next, using a Mathematica module created by a member of our research group, we have used this log-log plot to find the best-fit slope of the line, m, and thus the constant scaling factor relating how the resistance changes as a function of channel length for carbon nanotube networks. We repeated this procedure for networks with different probabilities of containing metallic wires in order to determine if the metallicity of the wires affects how the resistance of the network scales with channel length.

Results

We find a trend that as the channel length increases, the resistance increases as well. In addition, as the probability of metallic wires present in the network increases, the scaling factor m increases at a larger constant rate with increasing channel length. Figures 4, 5, and 6 show results from our computer simulation that allow us to determine how the resistance of carbon nanotube networks scales with channel length, depending on the amount of metallic wires present in the network. For further clarification, our results are summarized in Table 1.

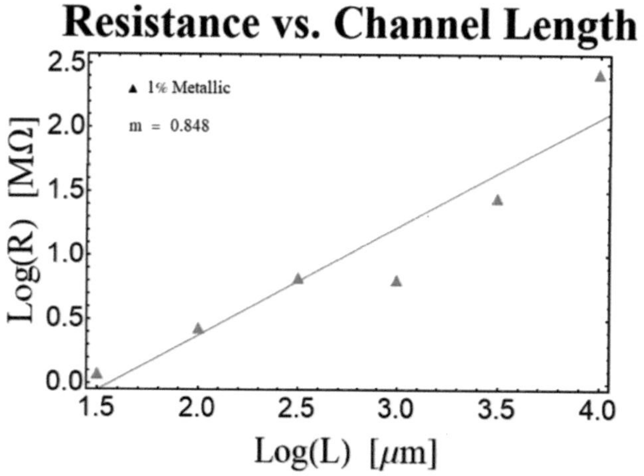

Figure 4: Resistance vs. Channel Length log-log plot. Blue line shows linear line of best-fit to the data, with the slope, m, equal to 0.848966. Channel length of networks ranges from 1.5 - 4 μm . Metallic wire probability is 1%.

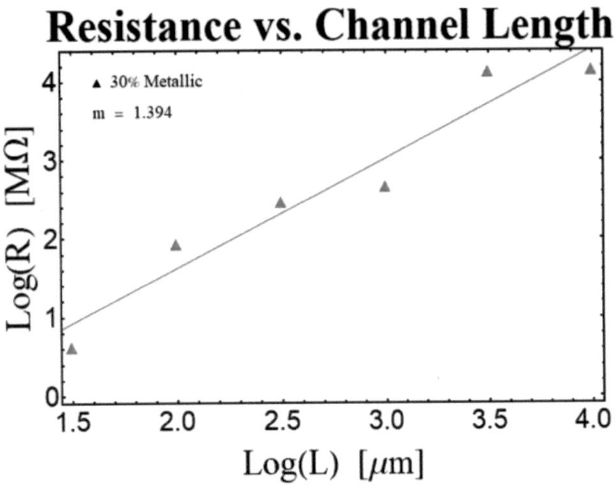

Figure 5: Resistance vs. Channel Length log-log plot. Blue line shows linear line of best-fit to the data, with the slope, *m*, equal to 1.39499. Channel length of networks ranges from 1.5 - 4 μ*m* . Metallic wire probability is 30%.

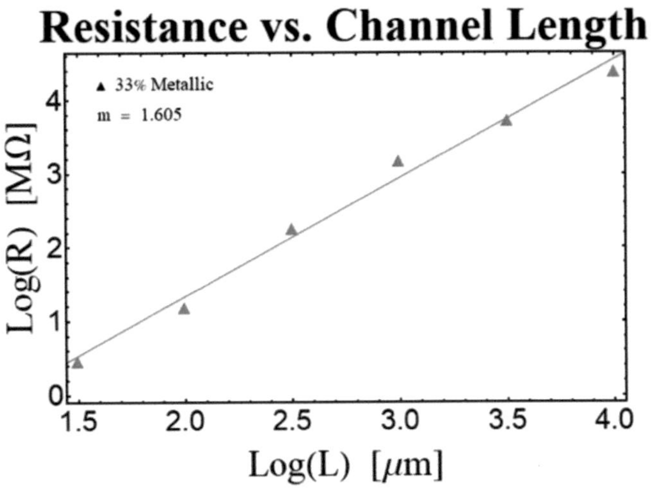

Figure 6: Resistance vs. Channel Length log-log plot. Blue line shows linear line of best-fit to the data, with the slope, *m*, equal to 1.60546. Channel length of networks ranges from 1.5 - 4 μ*m* . Metallic wire probability is 33%.

Table 1: Results of constant scaling factor *m* for various metallic wire probabilities.

Channel Length Range [m]	Density [tubes/ m^2]	Metallic Wire Probability	*m*
1.5- 4.0	1.6	1%	0.848966
1.5 - 4.0	1.6	30%	1.39499
1.5 - 4.0	1.6	33%	1.60546

Conclusion

In order to help inform the applicability of carbon nanotube networks for use as field effect transistors or as biosensors, we have run computer simulations in order to characterize the electrical properties of these networks, and specifically, how the resistance of these networks scales with the channel length for various amounts of metallic wires present in the networks. Other sources, using experimental or simulated networks, report values for the scaling constant *m* ranging from 1.2 -1.6 [4,5].

By furthering our understanding of resistances for carbon nanotube networks using computer simulations, we will be better able to determine how the resistance changes in the presence of charged molecules, such as potassium ions, allowing these networks to find applications as biosensors. Already, CNTs have been functionalized with aptamers that can bind molecules such as potassium [14]. Further investigations of the electrical properties of carbon nanotube networks will better characterize how these networks will behave as biosensors in the future.

Moving forward, we will continue utilizing computer simulations in order to determine how the resistance of carbon nanotube networks changes as the channel length of the networks is changed, for various amounts of metallic wires present in the networks. As yields of semiconducting tubes improve and carbon nanotubes find even more applications, our computer simulations will be able to offer more insight into how the electrical properties of these networks behave. We will also start to add an external electric field to our computer model in order to characterize the response of carbon nanotube networks to this field, which is important for the application of these networks as biosensors.

Acknowledgments

We would like to acknowledge Tyler Hooker for his help in our investigations of carbon nanotube networks.

References

Sumio Iijima, "Helical microtubules of graphitic carbon", Nature, 354, p 56–58 (1991)

K. S. Ibrahim, "Carbon nanotubes-properties and applications: a review", Carbon letters, 14 (3) Korean Carbon Society. Doi: 10.5714/CL.2013.14.3.131 (2013).

A J Page, F Ding, S Irle and K Morokuma, "Reports on Progress in Physics", 78, 3, (2015).

E. S. Snow, J. P. Novak, P. M. Campbell, and D. Park "Random networks ofW carbon nanotubes as an electronic material", Appl. Phys. Lett. 82, 2145 (2003).

V. K. Sangwan, A. Behnam, V. W. Ballarotto, M. S. Fuhrer, Ant Ural, E. D. Williams, "Optimizing transistor performance of percolating carbon nanotube networks", Appl. Phys. Lett. 97, 043111 (2010).

Ning Yang, Xianping Chena, Tianling Ren, Ping Zhang, Daoguo Yang, "Carbon nanotube based biosensors", Sensors and Actuators B: Chemical, 207, Part A, pp 690-715, (2015).

M. S. Fuhrer, J. Nygård, L. Shih, M. Forero, Young-Gui Yoon, Mazzoni, Jisoon Ihm, Steven G. Louie, A. Zettl,Paul L. McEuen, "Crossed Nanotube Junctions", 288, 5465, pp. 494-497, Science, (2000).

M. Stadermann, S. J. Papadakis, M. R. Falvo, J. Novak, E. Snow, Q. Fu, J. Liu, Y. Fridman, J. J. Boland, R. Superfine, and S. Washburn, "Nanoscale study of conduction through carbon nanotube networks", Phys. Rev. B, 69, 201402, (2004).

Masahiko Ishida, and Fumiyuki Nihey, "Estimating the yield and characteristics of random network carbon nanotube transistors", Appl. Phys. Lett., 92, 163507 (2008).

F. Bottacchi et al., "Nanoscale Charge Percolation Analysis in Polymer-Sorted (7,5) Single-Walled Carbon Nanotube Networks" Smal,l 12,31, 4211-4221 (2016).

C. Kocabas, N. Pimparkar, O. Yesilyurt S. J. Kang, M. A. Alam and J. A. Rogers, "Aligned Arrays of Single-Walled Carbon Nanotubes in Thin Film Type Transistors", Nano Lett., 7, 5, pp 1195–1202, (2007).

S. Kumar, J. Y. Murthy, and M. A. Alam, "Percolating Conduction in Finite Nanotube Networks", Phys. Rev. Lett. 95, 066802, (2005).

Ho-Kyun Jang, Jun Hee Choi, Do-Hyun Kim, Gyu Tae Kim, "On/off ratio enhancement in single-walled carbon nanotube field-effect transistor by controlling network density via sonication" , Applied Surface Science, 444, pp 442–447 (2018).

Han Yue Zheng, Omar A Alsager, Bicheng Zhu, Jadranka Travas-Sejdic, Justin Hodgkiss, Natalie Plank, "Electrostatic gating in carbon nanotube aptasensors " Nanoscale,8, 28, pp13659-68,(2016).

Percolation in Carbon Nanotube Networks

William P. Grieder (Humboldt State University), Will Gannett (Humboldt State University), Ruth Saunders (Humboldt State University)

Abstract

Carbon nanotubes (CNTs) have become increasingly useful in different applications since they were discovered in 1991 by Sumio Iijima [1]. One of their many useful qualities is their electronic properties [2]. These CNTs, when formed into a network, can be used as transistors [3] or biosensors [4]. Transistors are devices that regulate either current flow or voltage and act as a switch; they are a crucial component of computers. Biosensors detect the presence of biomolecules. Efficient transistors and biosensors already exist; however, they are expensive to manufacture compared to these CNT networks. The ability of the CNT networks to be transistors or biosensors relies on the percolation properties of the networks. As long as these networks percolate, current can pass through the network from a source electrode to a drain electrode, which can then be modulated by an electrical or biological signal to make devices such as transistors useful.

Keywords: carbon nanotube, nano, nano networks, percolation, percolation theory

Introduction

Carbon Nanotubes are formed when sheets of graphene (carbon arranged in hexagonal structure) roll up to make a tube. Depending on the way that the graphene sheet rolls up, CNT's can be either metallic or semiconducting, as shown in Figure 1. When CNTs form, roughly 33% are metallicand the rest are semiconducting. This is determined by the placement of the electrons relative to each other. It is important to minimize the presence of metallic CNT's in the network, to make it easier to utilize these networks as transistors or biosensors, because metal-

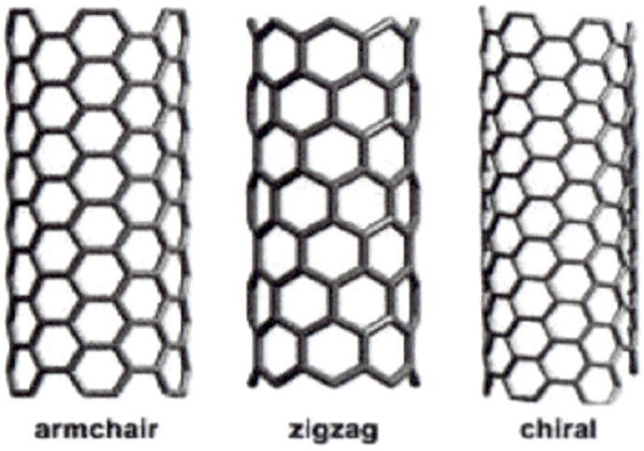

Figure 1: Diagram of Carbon Nanotubes showing different configurations: Armchair - Metallic, Zigzag and Chiral - Semiconducting [2]

lic tubes do not respond to external stimuli the same way semiconducting tubes do. Advancements in manipulating these CNT networks have made it possible for most metallic tubes to be stripped away, leaving only ~2% of the CNTs in the network as metallic [5].

Percolation

Percolation occurs when there is a pathway for something to travel through a given space, such as a forest fire traveling through a forest or using logs or rocks to cross a river [6]: only when there are enough connections can the fire spread through the forest or a person cross the river. In this case, percolation occurs when there is a pathway of CNTs that connects the source electrode to the drain electrode, allowing current to flow. In percolation theory, there is a useful value known as the percolation threshold. The percolation threshold is the lowest density that percolation occurs in an infinite network [6]. In finite networks, this is defined as the density where 50% of networks formed with that density will percolate. The percolation threshold of randomly placed sticks is ~6 sticks per square micron for 1 micron sticks, given by the Equation 1 [6].

$$\rho_{th}=4.26/(\pi l^2)$$

where ρ_{th} is the percolation threshold and l is the length of the sticks

(Equation 1)

It is crucial to fully understand the behavior of these networks in order to make reproducible devices. There is a balance between maximizing pathways for the current by having a high density and losing some of the sensitivity of the networks by being too far above the percolation threshold. The study of the percolation threshold for these networks will deepen our understanding of the behavior of these CNT network devices and allow us to provide experimentalists with information about optimal densities.

CNT Network Properties

Semiconducting materials are essential for transistors. Transistors behave like a switch turning on and off in response to an external electric field. Unlike metals, semiconductors have a resistance that can be changed when they are placed in an external electric field [7]. Most of the tubes in CNT networks are semiconducting and the relevant resistance for these networks are in the junctions between tubes. There are three types of junction: metallic-metallic (M-M), semiconducting-semiconducting (S-S) and metallic-semiconducting (M-S). The most responsive of these is the resistance between M-S junctions, which have a resistance on the order of 10^7 Ohms. The S-S junctions have a resistance on the order of 10^6 Ohms and M-M junctions have a resistance on the order of 10^5 Ohms [8]. For the network to function correctly, there cannot be a completely metallic percolation as the resistance of the network will not respond to the external field. When a percolating network is in an external electric field the resistance changes, allowing the network to have on and off states and act as a field effect transistor. The M-S junctions optimize the effectiveness of the transistor but also risk the possibility of metallic pathways [3]. Biosensors act in a similar way, where the electric field from charged particles affect the resistance of the network. The particle could then be determined from how the network resistance changes [4]. In order to determine the percolation threshold, which reduces the risk of metallic pathways, the percolation of these networks are simulated in a coding software, Python, for different network sizes and nanotube lengths. The output from the simulations are then compared to experimental results that show a percolation threshold of ~11 CNTs per square micron [9].

Method

The simulation sets an area for production of CNTs. The channel length is the distance between the electrodes where the width remains constant at 10 microns. The program then creates a stick to represent a CNT. An initial point is established within the channel, then another point is established at some specified distance (CNT length) and positioned at a random angle. The placement of these points are given by Equations 2 and 3, shown in Figure 2. These CNTs are giving a 2% chance of becoming metallic, while the rest become semiconducting. This production of sticks continues until there is a specified density of CNTs per square micron within the channel as shown in Figure 3.

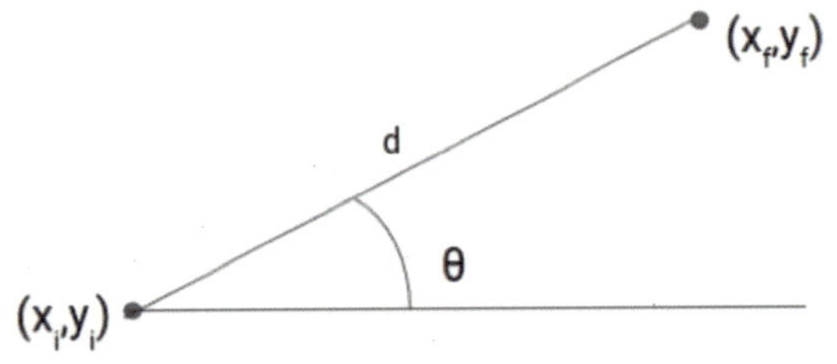

Figure 2:This stick represents a CNT where d is the length and θ is the random angle it is oriented in.

$xf = xi + d\cos(\theta)$ (Equation 2)
$yf = yi + d\sin(\theta)$ (Equation 3)

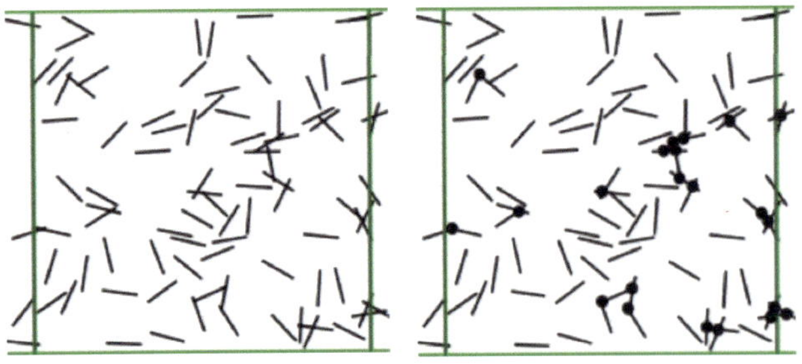

a) b)

Figure 3:
a. An example of a completed network
b. Same network with the junctions marked

Then the program analyzes the network to determine which CNTs are connected. The code stores the information of these connections in an adjacency matrix. Each component in the adjacency matrix is weighted with the value of conductance, given by Equation 4, between the respective junctions.

$$G = \frac{1}{R}$$

where G is conductance and R is resistance

(Equation 4)

The conductance of each junction is then used to find the conductance of the whole network. If there is zero conductance, then the network does not percolate between the electrodes. For a specific channel length, the simulation is run 100 times for increasing density. The fraction of the 100 networks that percolated is plotted as a function of density, as shown in Figure 4.

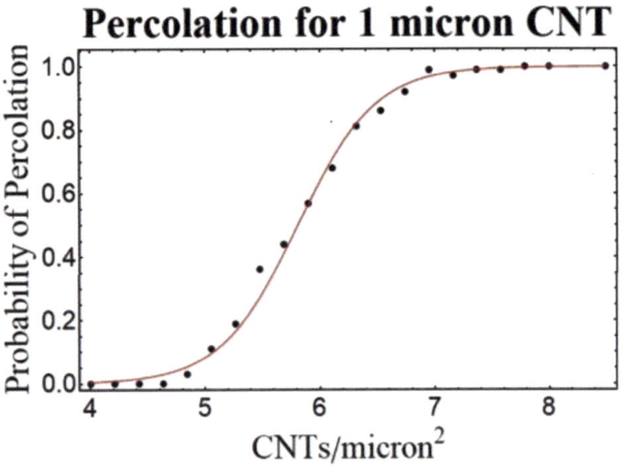

Figure 4: This plot shows the percolation progression for 10 micron channel length with 1 micron CNTs, where each point in this plot represents data from specific densities. (Density is in CNTs per square micron , Fraction of percolating networks out of the 100 simulated networks)

Results

This simulation is run for channel lengths ranging from 5 microns to 70 microns with a constant width of 10 microns. It is also run for CNTs of lengths of 1 and 2 microns, as shown in Figure 5. In reality, the network is made up of tubes with lengths distributed between 1 and 2 microns. As the length of the of the CNTs were increased, the density of CNTs per square micron needed to achieve percolation quickly decreased. As the channel lengths increased, the density needed for percolation increased. The percolation threshold (the density when ~50% of the networks produced at that density will percolate) for networks with 1 micron long CNTs, are between 5 and 8 CNTs per square micron, while networks with 2 micron long CNTs needed densities between 1 and 2 CNTs per square micron, as shown in Figure 6. These results come from a simulation that uses any angle to orient the CNTs, but the CNTs tend to naturally align with each other, giving them a higher probability of being within certain angles, which is simulated as shown in Figure 7. This may explain why the above results disagree with the experimental percolation threshold. The needed density for percolation increases as the CNTs become more aligned, as shown in Figure 8. This happens since they are less likely to connect with other CNTs when they are parallel, as shown in Figure 7.

Figure 5: This figure shows a 10 micron channel length with density of ~ 3 CNTs per square micron using (a) 1 micron CNTs and (b) 2 micron CNTs. Red CNTs are metallic, black CNTs are semiconducting.

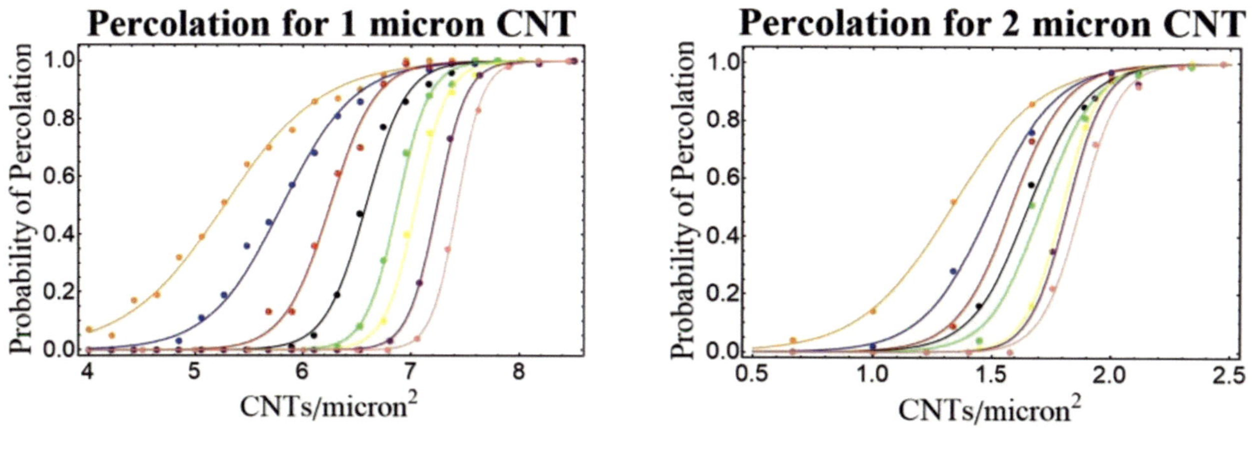

Figure 6: These plots show the progression of percolation as the channel length increases from 5 microns to 70 microns for (a) 1 micron CNTs and (b) 2 micron CNTs.

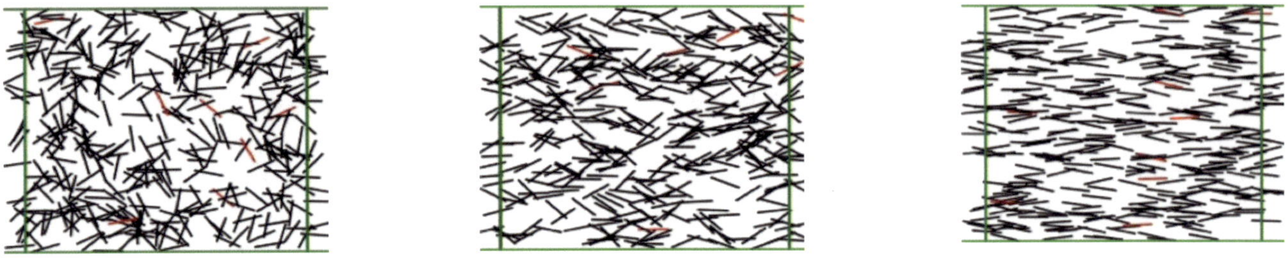

a) b) c)

Figure 7: The CNTs are oriented between (a) all angles, (b) -π/4 to π/4 radians, and c) -2π/15 to 2π/15 radians.

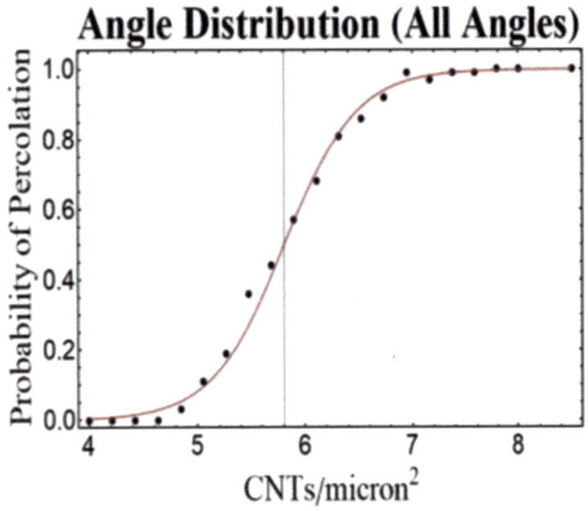

a)

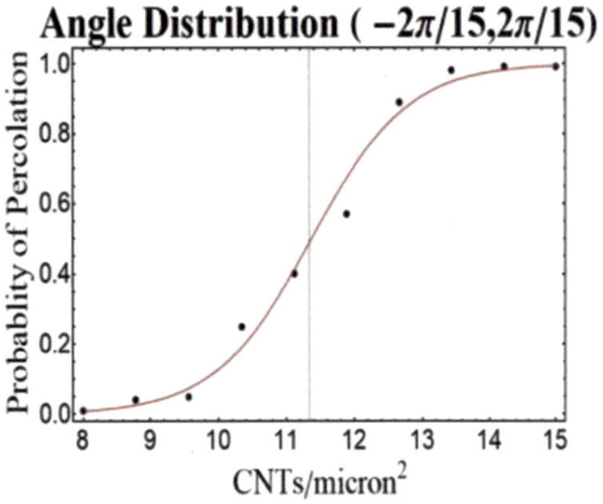

c)

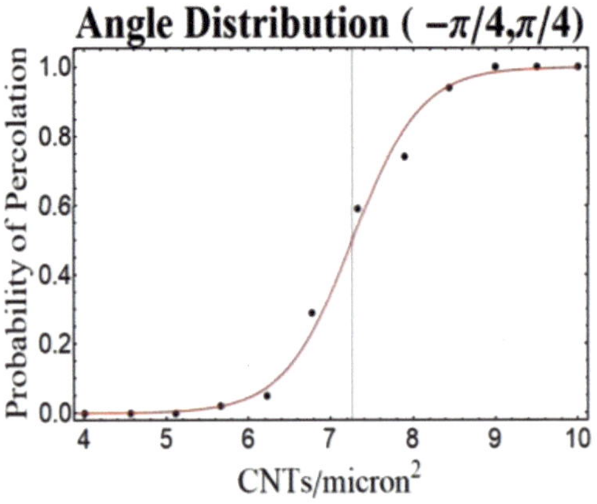

b)

Figure 8: Progression of percolation for 1 micron CNTs in 10 micron channel length (a) with all angles included, (b) with only angles between -π/4 to π/4 radians included, (c) with only angles between -2π/15 to 2π/15 radians included.

Future Work

Length Distribution
This code has been making networks with CNTs of a specific length of either 1 or 2 microns, when in the real nanotube networks, there are CNTs with lengths ranging from 1 to 2 microns long. Modeling this length distribution will affect the percolation threshold but will hopefully help bring the percolation threshold to the desired density.

Better Angle Distribution
Although there has been some angle distribution in the code, it doesn't include a distribution of probability for certain angles. The CNTs are much more likely to be parallel than they are likely to be at an angle to each other. Not only does this need to be changed but, since CNTs are so small, they don't have much effect on CNTs far away from them. This leads to localized groups of CNTs that could be aligned in one direction, while another localized group is aligned in another.

Curved CNTs
This code has been treating all the CNTs as though they are completely straight lines. Realistically, CNTs are curved, which might not have much effect on the percolation, but it is noticeable enough phenomena to include in the code.

Bundled CNTs
The same phenomena that causes the CNTs to align also causes them to bundle up together, which will greatly affect the percolation threshold of the networks. These bundles also tend to be mostly metallic, which will also greatly affect their use as transistors or biosensors.

Conclusion

The theory for percolation threshold for randomly placed sticks give a result of ~ 6 CNTs per square micron for a network of 1 micron CNTs [6], but our simulations produce results that still do not align with experimental results found in a previous study [9]. Without restricting the angles at which the CNTs lie, these simulations produced percolation thresholds at 5 - 7.5 CNTs per square micron for 1 micron CNTs and 1.25 - 2 CNTs per square micron for 2 micron CNTs. As the angles became more and more restricted, the percolation threshold increased much closer to the experimental results. However, in order to make the simulation represent real world CNT networks more accurately, some more modifications will need to be done to the code.

References

Iijima, Sumio (7 November 1991). "Helical microtubules of graphitic carbon". *Nature*. 354 (6348): 56–58.

A J Page, F Ding, S Irle and K Morokuma, "Reports on Progress in Physics", 78, 3, (2015).

V.K. Sangwan, A. Behnam, V.W. Ballarotto, M.S. Fuhrer, Ant Ural, E.D. Williams. Optimizing Transistor Performance of Percolating Carbon Nanotube Networks. Applied Physics Letters 97, 043111 (2010).

Ning Yang, Xianping Chena, Tianling Ren, Ping Zhang, Daoguo Yang, "Carbon nanotube based biosensors", Sensors and Actuators B: Chemical, 207, Part A, pp 690-715, (2015).

F. Bottacchi et al., "Nanoscale Charge Percolation Analysis in Polymer-Sorted (7,5) Single-Walled Carbon Nanotube Networks" Small 12,31, 4211-4221 (2016).

Stauffer, Dietrich. *Introduction to Percolation Theory.* London & Philadelphia, Taylor & Francis, (1985).

Neamen, Donald. "Semiconductor Physics and Devices". Elizabeth A. Jones.

M. Ishida and F. Nihey, "Estimating the yield and characteristics of random network carbon nanotube transistors," Applied Physics Letters 92, 163507 (2008).

H.Y. Zheng, N.O.V. Plank, " Facile fabrication of carbon nanotube network thin film transistors for device platforms", Journal of Nanotechnology, 14, 1-6, (2017).

J. Li and S.-L. Zhang, "Finite-size scaling in stick percolation", Physical Review E 80, 040104 (2009).

Made in United States
Troutdale, OR
11/03/2026

56844909R00050